THE GUINNESS BOOK OF

NUMBERS

THE GUINNESS BOOK OF

NUMBERS

ADRIAN ROOM

GUINNESS BOOKS

Project Editor: Honor Head
Editor: Anne Marshall
Design and Layout: Michael Morey

Published in Great Britain by Guinness Publishing Ltd,
33 London Road, Enfield, Middlesex

Typeset in Itek Rockwell Light
by Ace Filmsetting Ltd, Frome, Somerset
Printed and bound in Great Britain by
Hazell Watson & Viney Ltd, Aylesbury

British Library Cataloguing in Publication Data

Room, Adrian
Guinness book of numbers.
1. Number
I. Title
513'.2

ISBN 0-85112-372-4

CONTENTS

I *know* numbers are beautiful. If they aren't beautiful, nothing is.

(Paul Erdös, quoted in the *Sunday Times Magazine*, 27 November 1988)

INTRODUCTION

Numbers are part and parcel of our everyday lives. We find them in what we read, and we constantly use them in our day-to-day speech. We continually cope with sizes, weights, measures, distances, prices, ages, dates, categories and references. We find numbers in addresses, telephones, radio and TV programmes, and car registrations. We look at the clock or our watch to see the time — and there are numbers again.

Numbers are not only functional, but are a real part of the English language. We talk about having **forty winks**, or a cat having **nine lives**, or being on **cloud nine**. Many famous literary quotations also contain numbers, such as the witches' 'When shall we three meet again' in *Macbeth* or WB Yeats's 'Nine bean rows will I have there'. And many familiar number phrases come from the Bible, such as 'forty days and forty nights' and 'threescore years and ten'.

Then there are all the numbers in the world of sport: the scores, the points, the numbers of individual players.

This book offers a wide range of information about numbers, from the origins of numbers and counting, to the many usages and areas where numbers occur. There are numbers to inform, numbers to ponder, and numbers to entertain, too.

In short, the reader is offered a real 'Book of Numbers'. And if you want to discover how the *biblical* 'Book of Numbers' got its own name, that is in here, too!

So if you've ever wondered where numbers came from, or why they appear in so many aspects of our lives, this book will help you figure it out.

1 BACK TO SQUARE ONE

THE ORIGIN OF COUNTING AND RECKONING

The concept of **number** arose in prehistoric times, although originally the numbers themselves would not have been understood abstractly, as they are today. This does not mean that primitive man could not count or reckon— he was almost certainly able to judge the quantity of individual members of a group, such as people or animals, if only by naming each one separately and noting when one or more were missing. But although he could tell whether a group of hunters was complete or not, or could point to the various lakes suitable for fishing, he was probably unable to visualize a common link between, say, 'three hunters' and 'three lakes'.

Experts on early languages, and anthropologists who have studied still existing primitive peoples, have proved that original methods of counting were very basic, including only the lowest values and culminating in a word that meant simply 'many', with a different word for different objects. We preserve a similar idea today when we talk of a 'crowd' of people, a 'herd' of animals or a 'handful' of sweets.

It is known, for example, that the Vedda, who inhabited the forests of Sri Lanka (formerly Ceylon), never spoke of 'three trees' or 'two bulls', but simply said 'trees', 'bulls', and in 1972 a tribe of cave-dwellers in southern Mindanao (Philippines) were unable to answer the question 'How many people are there in your tribe?' although they could list individually all 24 members of their group.

The earliest numbers would thus have been 'one', 'two', 'many', with 'one' and 'two' arising in this order. 'One' was understood as the most basic number of all, and was indicated by a word for an object of which only one was seen to exist, such as 'Sun', 'Moon', 'I'. 'Two' would have been expressed by words naming parts of a human or an animal, such as 'eyes', 'ears', 'legs', 'wings'. If a thing was said to be 'eyes' it was thus 'as many in number as I have eyes'.

At the same time, 'two' was certainly regarded as a number differing from 'one' in concept in the same way that a person spoken to was different from the speaker. Seen this way, 'one' was thus 'I', 'two' was 'you', and anything else or anyone else ('he', 'she', 'it') would have been 'three'.

This radical concept lies behind the three grammatical categories of **singular** (indicating one of a thing), **dual** (indicating two) and **plural** (indicating many). The dual number existed in Greek and Russian, for example, and is still found today in some modern languages, such as Arabic, where *bait*, for instance, means 'house' and *baiten* means 'two houses'.

The earliest counting system based on 'one' and 'two' was thus a **binary** system, and the system is found even today for some primitive peoples, such as a group of South Sea Islanders who count beyond two by simply repeating the words for 'one' and 'two'. For them, urapun is 1, okosa is 2, okosa urapun 3, okosa okosa 4, okosa okosa urapun 5, okosa okosa okosa 6, and so on.

The modern binary system used for computer calculation is thus very old in origin.

As mankind developed, there was naturally an increasing need for numbers higher than 'three'. And just as 'eyes' had been used to express 'two', so other parts of the body were exploited for more advanced counting.

The most obvious units to use were the **fingers**. 'Four' could be expressed by the four fingers (omitting the distinctively different thumb), with the breadth of the four knuckles used to indicate a particular measurement. Even today 'hand' is used to express the height of a horse, with one hand equal to 4 in (10 cm), or an approximate measurement across the knuckles.

It is clear that 'four' was regarded as an important 'step' number beyond 'three', and traces of this notion still exist today in some languages. English, for example, retains 'once', 'twice', 'thrice' but nothing higher in this form, while Russian uses a form of the former dual (now the genitive singular) after 2, 3 and 4, but the plural after 5 and upwards.

However, it was all five fingers, including the thumb, that soon came to express 'five', while one hand could be added to the other to express 'ten'. This was the origin of the familiar **decimal** system of counting, with ten as a base.

Later, the ten **toes** could be brought into play to express higher numbers up to 20, so giving a **vigesimal** system, with base 20. This system is still found in French, for example, where the word for 80 is *quatre-vingts* ('four twenties').

It would have taken thousands of years, however, to evolve such a high level of reckoning, and it seems likely that the upper limit of counting, above 5, came about only gradually.

Some numbers below 10 were clearly regarded as 'step' numbers just as 'three' and 'five' had been. 'Seven', for example, was long thought of as an equivalent of 'many', and even in comparatively recent times tribes have been found where no numbers higher than 'seven' have been used. The German ethnologist Karl von den Steinen, for instance, described how in the 1880s he came across a South American Indian tribe with highly restricted counting words. He repeatedly made them try to count 10 grains of corn. They counted 'slowly but correctly to six, but when it came to the seventh grain and the eighth, they grew tense and uneasy, at first yawning and complaining of a headache, then finally avoiding the question altogether or simply walking off'.

The frequent occurrence of 'seven' today in a number of common phrases, such as 'seven seas' and 'seven deadly sins', almost certainly derives from the primitive concept of 'seven' as 'many'.

'Nine', too, was often regarded as a 'step' number after 'eight' (itself twice 'step' number 'four'), and there is sound evidence for linking the words 'nine' and 'new' in many languages. In modern French, the word *neuf* has both meanings still.

In the course of time, the **abstract** concept of number evolved, and humans began comparing a quantity of identical or similar objects with one such object, which itself was the standard. One such standard unit was the finger, as mentioned. But a need arose for a record of counted objects to be kept. The answer lay in the **tally stick**, where the unit of counting was not a finger but a notch, made on a stick or piece of wood with a sharp instrument such as a knife or axe.

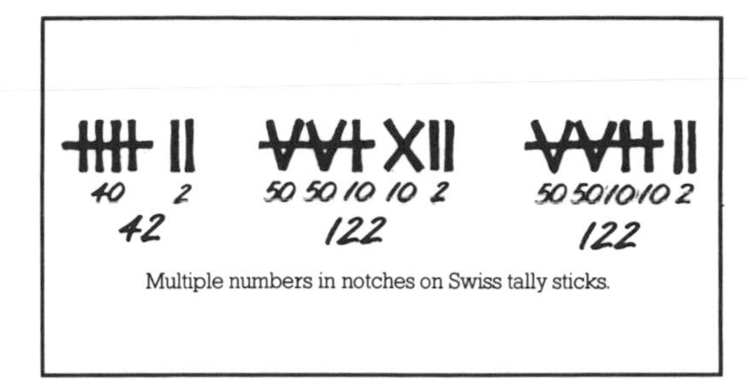

Multiple numbers in notches on Swiss tally sticks.

One such notch indicated '1', two notches meant '2', and so on. For higher numbers, a distinctive longer notch was used to indicate a multiple, or else a particular number of notches, usually 10 (for the 10 fingers), was scored through to denote a group of 10, thus: ++++++++++.

This system of calculation lies behind the modern English word 'score' to mean either '20' (i.e. 10 hands plus 10 toes) or 'total number of points', as in a game or sport. In the earlier days of cricket, the actual word 'notch' was used similarly for 'run', and a text of 1737, for example, tells how 'Kent side went in first and got 99 notches'. A hundred years later, Dickens's *Pickwick Papers* recounts how 'All-Muggleton had notched some fifty-four'.

Just as traces of the vigesimal system are still found today, so a **sexagesimal** or 60-based system of counting remains in the 60 minutes that we have to the hour and 60 seconds to the minute. This method was adopted by the Babylonians in the earliest times, while the Sumerians, who lived in the fourth millennium BC, subsequently evolved their own numeral notation based on gradations of 60, as well as the established decimal system.

A sexagesimal system implies one based on 5 units of 12, and the Babylonians also favoured the **duodecimal** or 12-based method of counting, as did the Chinese and the Romans.

The duodecimal reckoning is still familiar today in the 12 hours that make a day or night, and the 12 months that make up a year, while British currency was duodecimal-based (12 pence to a shilling, 240 pence to a pound) until as recently as 1971, when decimal coinage was introduced.

Although the decimal system is more natural, thanks to the 10 fingers (on which many young children still count), the duodecimal has an advantage, especially when calculating, for 12

FINGERS COUNT

In his novel *The Wind Cannot Read*, published in 1947, Richard Mason tells how Sabby, a young Japanese woman in India, which was then at war with Japan, was introduced as Chinese to an Englishman who had been living a long time in India and who knew the ways of the East.

'This is Miss Wei,' I said.

'Really?' He extended his neck to look at Sabby closely, as though he were short-sighted. His eyes were screwed up quizzically. 'Nonsense,' he said. 'Count on your fingers. Count up to five.'

Sabby looked bewildered; she did not know whether this curious man was being funny or savage. She held up a doubtful hand.

'That's right. Count. One, two, three, four, five.'

'One, two, three, four, five,' Sabby said uncertainly.

Mr Headley burst out in delight. 'There you are. Did you see? Did you see the way she did it? Started with an open palm and closed the fingers one by one. Ever seen a Chinese do that? Of course not. They count like the English, starting with a closed fist. Japanese!' he diagnosed triumphantly.

will divide by 2, 3, 4 and 6, but 10 will divide only by 2 and 5. For this reason, 'twelve' came to have its alternative designation of 'dozen', just as 'twenty' was also 'score', and both these words were still used in English to mean simply 'many', much as 'seven' had been regarded earlier. We can still say to a mischievous child 'If I've told you once, I've told you a dozen times', and, when we are busy, say 'I've scores of things to do'.

When 'eleven' and 'twelve' were originally established, they were both also regarded as 'step' numbers, like 'four' and 'nine', and a modern legacy of this survives in the actual words for these numbers in some languages today, such as English 'eleven' and 'twelve' themselves, which do not involve a mention of 'ten' (as in 'thirteen'). German *elf* and *zwölf* are similar.

THE EMERGENCE OF WRITTEN NUMERALS

With the development of writing, so the ways of expressing numbers in a convenient **visual** form evolved. To begin with, the lowest numbers (1, 2, 3) were represented on the writing material (papyrus, bark, clay tablets and the like) by simple lines or dashes, much like the basic notches for these same numbers on the tally stick. Even today, the Chinese numerals 1, 2 and 3 are denoted by the appropriate number of strokes, and we are all familiar with the similar I, II and III of Roman numerals.

Overall, the introduction of separate symbols for individual numbers was a tremendous advance. We now take it for granted, and have become accustomed to a notational system where a particular quantity can be expressed either by a word ('seven') or its equivalent symbol ('7'). The symbols themselves, especially the so-called Arabic figures, have now become virtually international, making for ease of communication. Even so, each language still needs its own word for the number (seven, *sept*, *sieben*, etc.), which means that most countries have two ways of denoting the same figure, one international, the other national.

For the Chinese and Japanese, however, there is no such duality of denotation, for the pictogram that depicts the number actually *is* that number, and Chinese *sān*, for example, is both the term for 'three' and the word for the pictogram (of three parallel strokes) that represents this number.

Such a system of numerical notation is unique, and is the only known instance in history in which the written number word has come to be the same as the numeral.

Today the Arabic figures are in common usage, and the Roman numerals are kept for more specialized uses, such as the figures on some clock and watch faces.

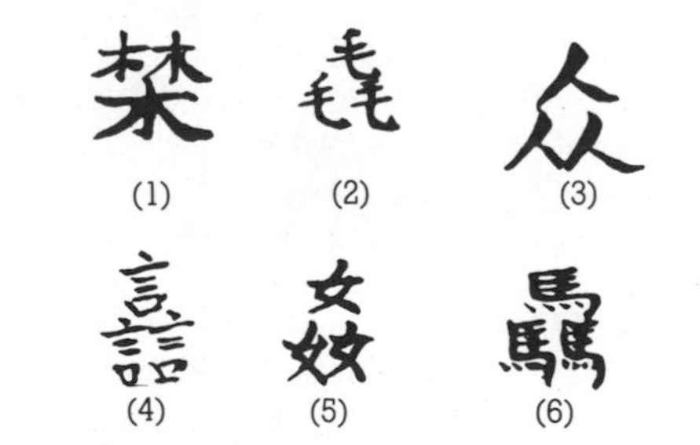

Three as the plural in Chinese: (1) forest = 3 × tree; (2) fur = 3 × hair; (3) all = 3 × man; (4) speak endlessly ("much") = 3 × speak (mouth from which words emerge); (5) gossip = 3 × woman; (6) gallop (ride "much") =3 × horse.

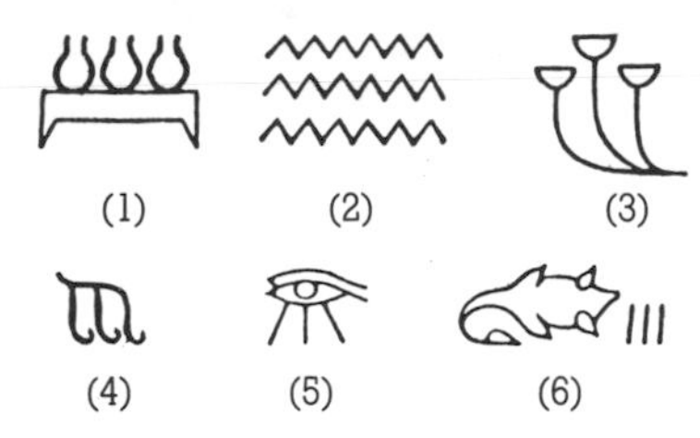

Three as the plural in Egyptian: (1) flood = heaven with 3 water jugs; (2) water = 3 × wave; (3) "many" plants = 3× plant; (4) hair = 3 hairs; (5) weep = eye with "many" (= 3) tears; (6) fear = dead goose with 3 vertical strokes, the general plural sign, next to the ideogram.

THE EVOLUTION OF NUMERATIONAL SYSTEMS

Arabic and Roman numerals are not the only types, however. One of the most ancient numerational systems was that of Egyptian **hieroglyphics**, using the stylized pictorial symbols that characterized this form of sacred writing. ('Hieroglyphics' means literally 'holy carving'.)

The script, which evolved over 5000 years ago, had short straight strokes, representing measuring sticks, for the numbers 1 to 9; a reversed 'U' figure, representing a hobble (for fettering the legs of a cow), for 10; a curled figure, representing a measuring rope, for 100; a sign representing a lotus flower for 1000; the pictogram of an index finger for 10,000; the figure of a tadpole (indicating high plurality) for 100,000; a representation of a man with his arms raised (as if expressing great surprise) for 1 million, and a picture of the sun to indicate 10 million. These different figures could then be combined as necessary.

Later, with the development of Egyptian culture, such hieroglyphic writing gave way to **hieratic**. This was a shorthand written version of hieroglyphics used by Egyptian priests (the term itself means literally 'priestly'), and evolved individual symbols for each multiple of 10, for example, instead of merely repeating the '10'

	10	20	30	40	50	60	70	80	90
Pictographic	∩	∩∩	∩∩∩	∩∩∩∩	∩∩∩∩∩	∩∩∩∩∩∩	∩∩∩∩∩∩∩	∩∩∩∩∩∩∩∩	∩∩∩∩∩∩∩∩∩
Hieratic	[illegible]	[illegible]	[illegible]	[illegible]	[illegible]	[illegible]	[illegible]	[illegible]	[illegible]

symbol. Hieratic was then in its turn succeeded by the **demotic** system, used by the ordinary literate class of people outside the priesthood. (The word literally means 'of the people'.) In this, symbols that represented letters of the alphabet were also utilized as numerals. The alphabetical symbols themselves were usually stylized representations of an object whose name began with that letter. For example, the symbol that came to stand for letter 'b' and figure '2' was a stylized drawing of a house, representing the initial sound of the Semitic word for 'house', which was *beth*. (Hence the name of the second letter of the Greek alphabet, beta, and even the word 'alphabet' itself, which represents the first two letters of the Greek alphabet, alpha and beta.)

The use of demotically-derived letters as numerals was also a general feature of Phoenician, Hebrew and Greek, although to adapt the alphabet to their own language, the Greeks altered some of the letters, which were originally consonants, to vowels. They also added three letters (phi, chi and psi).

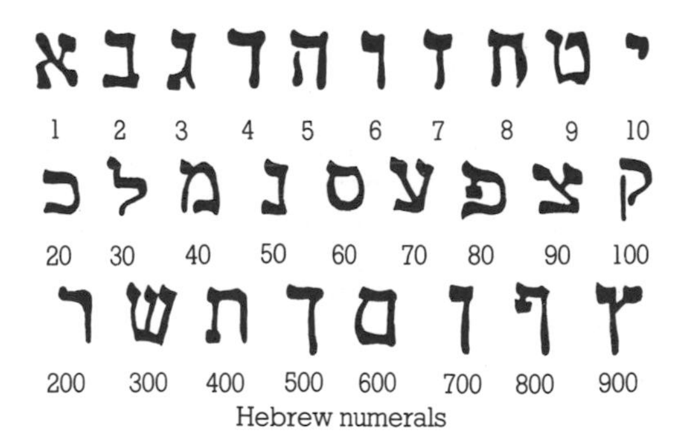

Hebrew numerals

Letters of the alphabet are still used in a numerical role today, with (a), (b), (c), for example, denoting the first, second and third points of a numbered paragraph or itemized statement. Equally, letters are used for various grading systems where numerals would also do. British nurses, for instance, are on a pay scale that runs from A (the lowest, for an unqualified auxiliary) to I (the highest, for a nursing manager). Letters are also used as numbers for certain measuring scales and sizes, especially where an object can be measured in two different ways. Footwear, for instance, is measured in length numerically, but in width by letter.

Not all writing systems, however, used letters as numerals. The Chinese, for example, used,

(a) 6 7 8 10 20 30 40 50 60 70 80 90

(b)

(a) numbers 6 to 90: (b) the number 46 431 in monogram form.

Chinese 'rod' numerals.

and still use, different numerical notations from the simple 'rod' system, based on straight strokes, to the modern 'common' numerals, 'official' numerals, which are highly-ornamented versions of these, designed to prevent fraud, and 'commercial' numerals, which are a simplified notation. In the past, too, the Maya and Aztec numerations were vigesimal, and used a variety of symbols, such as dots, lines and circles, together with devices that were more obviously pictographic.

Most of these early numerical systems lacked a zero, although the Maya notation had one, in the form of a shell, and current Chinese 'commercial' numerals also include a zero.

The familiar Roman numerals, however, had no zero, unlike the much more familiar Arabic numerals, which were subsequently introduced to Europe.

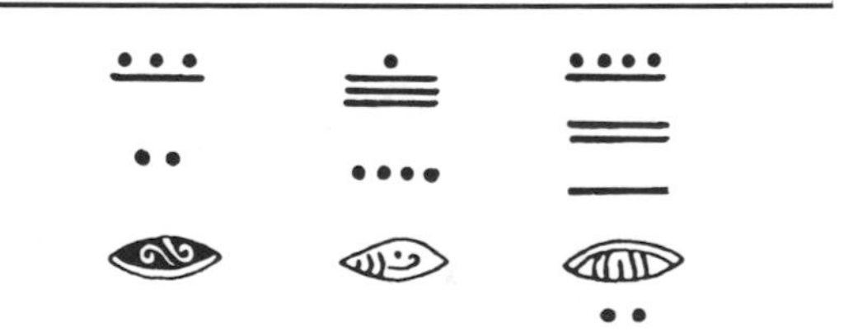

Mayan abstract place-value notation containing the oldest zero in the New World. The zero sign looks like a snail's shell. The units are the early numerals with a quinary grouping, and there are no numerical ranks. The three Mayan numbers illustrated, from top to bottom, are as follows:
left: Maya $\bar{8}20 = 8 \times (20 \times 18) + 2 \times 20 + c$ = Indian 2920;
centre: Maya (16)40 = $16 \times (20 \times 18) + 4 \times 20 + 0$ = Indian 5840;
right: Maya 9(10)502 = $9 \times (20^3 \times 18) + 10 \times (20^2 \times 18) + 5 \times (20 \times 18) + 0.20 + 2$ = Indian 1 369 802.

THE ROMAN NUMERALS

As we now know and use it, the **Roman** numerical system consists of seven basic integers, each denoted by a particular letter: I for 1, V for 5, X for 10, L for 50, C for 100, D for 500 and M for 1000. The system is a decimal one, like the Arabic notation, but unlike the latter it is non-positional, that is, its individual letters do not indicate the units, tens, hundreds or thousands as the Arabic numerals do. In the Arabic notation, the first figure of the number 4862, for example, indicates the thousands (4), the second shows the hundreds (8), the third gives the tens (6) and the final figure indicates the units (2). In Roman numerals, the same number would be MMMMDCCCLXII, where it needs the four Ms to give the number of thousands, the D (500) and the three Cs (300) to give the hundreds when added together, the L (50) and the X to give the tens when similarly added, and the final two Is to give the units.

Roman numerals almost certainly developed from existing Etruscan symbols, used in central Italy in the region that approximates to modern Tuscany. These basically consisted of individual vertical strokes to indicate 1 to 4, an inverted 'V' for 5, a cross (X) for 10, and combinations of these for higher figures. The strokes up to 4 can be regarded as direct representations of the original four fingers of the hand, as notched on early tally sticks. It is probably significant that all the Roman numerals below C (100) consist of straight lines, suitable for easy writing or engraving.

If the numerals I, II, III and IIII represent the four fingers, then V may well represent the open hand, as formed by the four fingers and thumb. X would then be two such 'hands' put back to back. But it is possible that V may equally represent half 'X', with X itself originating simply from a single stroke that had been crossed through (ł), much as the ten notches on the tally stick were crossed through to show a multiple unit.

The 'L' that now means 50 probably developed from an early form of the Greek letter chi, with a form such as ᴪ gradually evolving to ⋔, then ⊥, then finally ∟. Otherwise it may have evolved from a halving of the Etruscan ✳ (100), perhaps via ↓, then ⊥, then ∟. The present 'L' does not represent the first letter of a Latin word.

The 'C' that now means 100 is only coincidentally the initial letter of Latin *centum*, 'hundred'.

The numeral probably developed either from an early form of the Greek letter 'theta', with a symbol such as ⊕ giving first ⊖, then ⊙, then Ͼ, then finally C, or else from an Etruscan numeral of similar type. Other theories derive C from the first half of the figure ↀ, with the right half giving 'D' (500), the next number up, or from the letter 'X' in parentheses, as (X).

The 'D' that is 500 does not represent the initial letter of a Latin word, but probably evolved as described above, that is, from the figure that may also have given the 'C' of 100.

The 'M' that is 1000 was only coincidentally influenced by the Latin word *mille*, 'thousand'. It seems to have evolved from an early form of the Greek letter 'phi', through stages such as ϕ, Φ, ↀ, and ψ, or else from an 'X' in parentheses (as for 'C', above). Alternative figures to denote 1000 have been recorded as (|), ∞, ⊠, and ∼.

GREEK NUMERALS

Somewhat similar to the system of Roman numerals was the so-called **Attic** or **Herodian** system of Greek numerals. 'Attic' refers to the Ancient Greek territory of Attica, whose capital was Athens, and 'Herodian' derives from the name of the 2nd century AD grammarian Aelius Herodianus, who briefly described the system in his writings.

The system arose some 700 years before the time of Herodianus, however, and was based on the initials of the Greek words for 5 (ΠΕΝΤΕ), 10 (ΔΕΚΑ), 100 (ΗΕΚΑΤΟΝ), 1000 (ΧΙΛΙΟΙ), and 10,000 (ΜΥRΙΟΙ), with single vertical strokes representing the units.

The system was both denary- and quinary-based (i.e. on both 10 and 5), with the quinary grouping indicated by a multiple expressed in the form of a small symbol inside the 'Γ' that represented 5. This gave 𐅄 as 50, 𐅅 as 500, 𐅆 as 5000 and 𐅇 as 50,000. Like the Roman numerals, the Greek system was cumbersome, especially for the higher values, so that 234 was written ΗΗΔΔΔΙΙΙΙ, for example, and 1989 was Χ𐅅ΗΗΗΗ𐅄ΔΔΔΓΙΙΙΙ.

Roman numerals were used widely in European countries, however, down to medieval times, and were even preferred to Arabic numerals for their reasonable security against fraud when stating sums of money. Any digit of the number 2381, for example, can be altered to cause a substantial difference (as much as 7000 if the 2 becomes a 9), but the Roman equivalent, MMCCCLXXXI, cannot be easily altered to produce a new number.

NUMBER 10 AND THE NATURAL WORLD

Annus erat, decimum cum luna receperat orbem,
Hic numerus magno tunc in honore fuit.
Seu quia tot digiti, perquos numerare solemus:
Seu quia bis quino femina mense parit:
Seu quod adusque decem numero crescente venitur,
Principium spatiis sumitur inde novis.

A year was past when the number of full moons was ten,
This number was held in great honour then:
Either because that is the number of fingers on which we reckon,
Or because a woman gives birth to her child in two times five months,
Or because the numbers increase up to ten,
And from there a new round of reckoning begins.

(Ovid, *Fasti* III, AD 1-8)

ARABIC NUMERALS

The now internationally common **Arabic** numerals were adopted by most European countries only in about 1500. They are not Arabic in origin at all, however, but Indian, and are only named 'Arabic' simply because Europeans learned about them through Arabic writers. Unfortunately, little is known about their origin, except that they evolved in India some time between the 2nd century BC and the 6th century AD. It was during this period that Indian mathematicians came to realize their advantage over the abacus for calculation purposes. It was only in the 9th century AD that the Arab mathematician Al-Khuwārizmi came to adopt them outside India for his calculations, and it is a corruption of his own name that has given the modern term 'algorism' to refer to the Arabic (i.e. Indian) decimal system of reckoning, and thus the term 'algorithm'.

The advantage of the Arabic numerals over the Roman can be instantly appreciated. Each number from 1 to 9 has its own symbol, with '0' used to express 'zero' (which the Roman system lacked). Any number, of any magnitude, can be expressed accurately and reasonably compactly by using these ten digits, so that 638,234, for example, means (reading from right to left) 4 units + 3 tens + 2 hundreds + 8 thousands + 3 tens of thousands + 6 hundreds of thousands. The number is instantly comprehensible, even though one digit (3) occurs twice (in different values). Moreover, the Roman system of numerals works on a subtractive system, with lower figures subtracted from a following higher one (such as XIX for 19, where the 'I' is subtracted from the 'X' to give the 9). Arabic numerals have a straightforward positional principle of addition or multiplication, so that '48', for instance, can be interpreted as either '8 units + 4 tens' or '8 + (10 × 4)'.

The basic cumbersome nature of Roman numerals for expressing high numbers caused them to be generally abandoned for scientific purposes in Europe from about 1600, and the introduction of logarithms in the 17th century, using Arabic, sounded their death knell.

Modern Arabic numerals, as used in the Arabic-speaking countries of the Middle East, such as Egypt and Saudi Arabia, evolved as an Eastern Arabic version of the original Indian numbers, as distinct from our own 'Arabic' numerals, which represent the Western Arabic

٠ ١ ٢ ٣ ٤ ٥ ٦ ٧ ٨ ٩   ١٩٥٥/١/٣١

0 1 2 3 4 5 6 7 8 9   1955·1·31

The Eastern Arabic (above) and our numerals, which derived from the Western Arabic version.

version. There are now few similarities between the two types, although '1' in each system is represented by a vertical stroke, and '7' remains essentially the same, except that the Eastern Arabic version of the number is differently angled to our Western Arabic figure. The other main difference is that the Eastern Arabic numbers contain more straight lines, whereas the Western Arabic figures are generally more rounded. Zero in each is similar, however, with a 'dot' in Eastern Arabic and a circle in Western.

It is noteworthy that although modern Arabic script runs from right to left, the numerals run from left to right, as ours do.

THE ABACUS

A calculating device that has served for many centuries as a halfway house between human fingers and the modern computer is the **abacus**. Its name derives from Greek *abax*, a term used for a board covered in sand on which calculations could be traced, and itself borrowed from Hebrew *ābhāq*, 'dust'.

The abacus is probably of Babylonian origin, but developed into two distinct types, eastern and western, from Roman times.

The **eastern** or Chinese abacus, the *hsuàn pán*, 'calculating board', is also used in Indo-China and Japan. It consists of a frame divided into two halves by a horizontal bar. The two beads in the upper section of the frame each represent five units, and the five beads below the bar represent one unit each, reading downwards (1 to 5). As the abacus operates on the decimal system, the far right-hand row of beads represents the units, the next is the tens, the third the hundreds, the fourth the thousands, and so on. When a simple addition is carried out, the beads are moved to the bar. The number shown on the abacus illustrated is 1930, with the '9' represented by one '5' bead and four '1' beads.

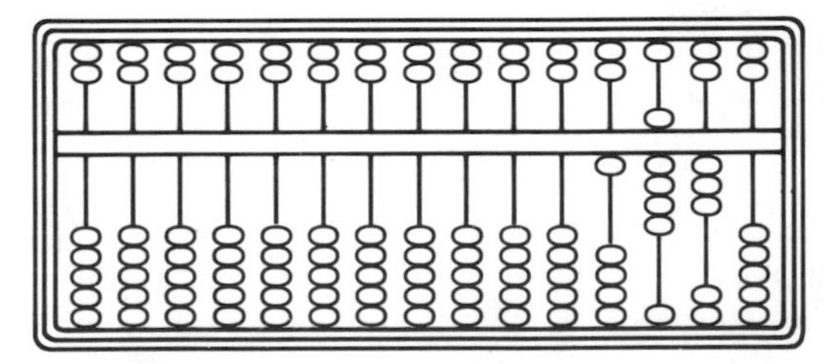

The **western** abacus, still in use in the Soviet Union, was introduced to Russia in the 16th century, and consists of a single frame with rows of ten beads, each representing a single unit, with the bottom row of beads working as units, the next row serving as the tens, the third up as the hundreds, and so on. For ease of rapid manipulation, the units 5 and 6 beads are always darker in colour.

To carry out a calculation, the beads are moved to the left of the frame. The number shown on the abacus pictured is 401.28, with the two bottom rows (below the four beads) expressing the decimal fraction, in tenths and hundredths. The four beads themselves are used to express a simple fraction in quarters.

THE BINARY SYSTEM

The modern **binary** system of reckoning, with a

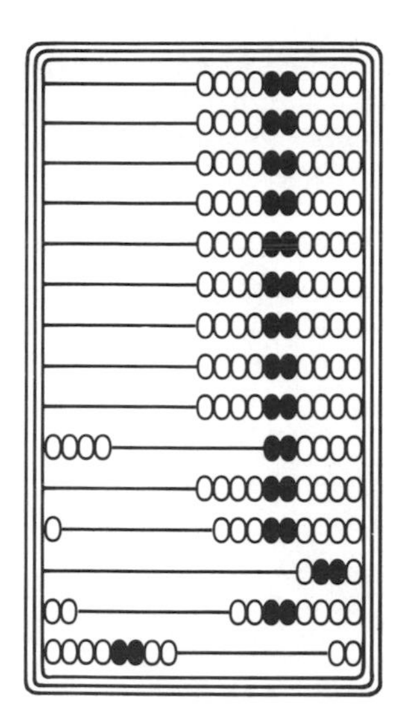

base of 2, is of ancient origin, as already mentioned. As used by computers, it became generally familiar from the 1960s. The system uses only the two symbols 0 and 1, each of which is known as a 'binary digit' or 'bit' (for short). These digits represent the electrical pulses that flow through a computer, with '0' standing for a nil or low voltage, and '1' representing a higher voltage. More simply, they can be thought of as the 'off' and 'on' positions of a switch.

Although '0' always equals zero, as in the familiar decimal system, '1' varies its value depending on its position in the number, and it doubles its value every time it moves one column to the left. Thus 0001 equals simply '1' in the decimal system, but 0010 equals '2', as the '1' has doubled by moving to the second column to the left. '3' is expressed by 0011 (i.e. by the equivalent of 2 + 1, or a combination of 0010 and 0001). This means that decimal '4' will be 0100 in binary, with the 1 moving a further column to the left, and doubling again in value.

It follows that the columns themselves, reading from right to left, represent decimal 1, 2, 4, 8 and

NUMERAL NOTATIONS OF DIFFERENT RACES

	Egyptian			Babylonian	Phoenician	Syriac	Palmyrene	Attic Greek	Roman	Chinese		Kharoshti	Nasik	Maya	Ionic Greek	Slavonic		Hebrew	Syriac	Arabic	Georgian	Armenian
	Hieroglyphic	Hieratic	Demotic							Common	Commercial					Cyrillic	Glagolitic					
0											O			[illegible]								
1	I	I	I	[illegible]	I	I	I	I	I	一	〡	I	—	•	α	а	Ⰰ	א	ܐ	ا	ა	Ա
2	II	[illegible]	[illegible]	[illegible]	II	[illegible]	II	II	II	二	〢	II	=	••	β	в	Ⰱ	ב	ܒ	ب	ბ	Բ
3	III	[illegible]	[illegible]	[illegible]	III	[illegible]	III	III	III	三	〣	III	≡	•••	γ	г	Ⰲ	ג	ܓ	ج	გ	Գ
4	IIII	[illegible]	[illegible]	[illegible]	[illegible]	[illegible]	IIII	IIII	IV	四	〤	X	[illegible]	••••	δ	д	Ⰳ	ד	ܕ	د	დ	Դ
5	[illegible]	[illegible]	[illegible]	[illegible]	[illegible]	[illegible]	y	Γ	V	五	〥	IX	[illegible]	—	ε	є	Ⰴ	ה	ܗ	ه	ე	Ե
6	[illegible]	[illegible]	[illegible]	[illegible]	[illegible]	[illegible]	'y	ΓΙ	VI	六	〦	IIX	[illegible]	[illegible]	ϛ	ѕ	Ⰵ	ו	ܘ	و	ვ	Զ
7	[illegible]	[illegible]	[illegible]	[illegible]	[illegible]	[illegible]	"y	ΓΙΙ	VII	七	〧		7	[illegible]	ζ	з	Ⰶ	ז	ܙ	ز	ზ	Է
8	[illegible]	[illegible]	[illegible]	[illegible]	[illegible]	[illegible]	"'y	ΓΙΙΙ	VIII	八	〨	XX	[illegible]	[illegible]	η	и	Ⰷ	ח	ܚ	ح	ჱ	Ը
9	[illegible]	[illegible]	[illegible]	[illegible]	[illegible]	[illegible]	""y	ΓΙΙΙΙ	IX	九	〩		?	[illegible]	θ	ѳ	Ⰸ	ט	ܛ	ط	თ	Թ
10	∩	Λ	[illegible]	[illegible]	—	7	⊃	Δ	X	十	十	7	[illegible]	=	ι	і	Ⰹ	י	ܝ	ى	ი	Ժ
15	∩ IIIII	[illegible]	[illegible]	[illegible]	[illegible]	[illegible]	'⊃	ΔΓ	XV	十五	十〥			≡	ιε	єі	[illegible]	יה	ܝܗ	يه	იე	ԺԵ
20	∩∩	[illegible]	[illegible]	[illegible]	[illegible]	O	3	ΔΔ	XX	二十	〢十	3	θ	[illegible]	κ	к	Ⰺ	כ	ܟ	ك	კ	Ի
30	∩∩∩	[illegible]	[illegible]	[illegible]	[illegible]	70	⊃3	ΔΔΔ	XXX	三十	〣十				λ	л	Ⰻ	ל	ܠ	ل	ლ	Լ
40	∩∩∩∩	[illegible]	[illegible]	[illegible]	[illegible]	OO	33	ΔΔΔΔ	XL	四十	〤十		X		μ	м	Ⰼ	מ	ܡ	م	მ	Խ
50	∩∩∩∩∩	[illegible]	[illegible]	[illegible]	[illegible]	700	⊃33	𐅄	L	五十	〥十	733			ν	н	Ⰽ	נ	ܢ	ن	ნ	Ծ
60	∩∩∩∩∩∩	[illegible]	[illegible]	[illegible]	[illegible]	OOO	333	𐅄Δ	LX	六十	〦十	333			ξ	ѯ	Ⰿ	ס	ܣ	س	ჲ	Կ
70	∩∩∩∩∩∩∩	[illegible]	[illegible]	[illegible]	[illegible]	7OOO	⊃333	𐅄ΔΔ	LXX	七十	〧十	7333	[illegible]		ο	о	Ⱀ	ע	ܥ	ع	ო	Հ
80	∩∩∩∩∩∩∩∩	[illegible]	[illegible]	[illegible]	[illegible]	OOOO	3333	𐅄ΔΔΔ	LXXX	八十	〨十				π	п	Ⱁ	פ	ܦ	ف	პ	Ձ
90	∩∩∩∩∩∩∩∩∩	[illegible]	[illegible]	[illegible]	[illegible]	7OOOO	⊃3333	𐅄ΔΔΔΔ	XC	九十	〩十				ϙ	ч	Ⱂ	צ	ܨ	ص	ჟ	Ղ
100	[illegible]	[illegible]	[illegible]	[illegible]	[illegible]	7'	⊃'	H	C	百	θ	λI	[illegible]		ρ	р	Ⱃ	ק	ܩ	ق	რ	Ճ
200															σ	с	Ⱄ	ר	ܪ	ر	ს	Մ
300															τ	т	Ⱅ	ש	ܫ	ش	ტ	Յ
400									CD						υ	у	Ⱆ	ת	ܬ	ت	ჳ	Ն
500	[illegible]	[illegible]	[illegible]	[illegible]	[illegible]		⊃'	𐅅	D	五百	〥百		[illegible]		φ	ф	Ⱇ	תק		ث	ფ	Շ
600									DC						χ	х	Ⱈ	תר		خ	ქ	Ո
700															ψ	ѱ	Ⱉ	תש		ذ	ღ	Չ
800															ω	ѡ	Ⱎ	תת		ض	ყ	Պ
900									CM						ϡ	ц	Ⱌ	ץ		ظ	შ	Ջ
1000	[illegible]	[illegible]	[illegible]	[illegible]			[illegible]	X	M	千	千		9		͵α	҂а	Ⱋ	א̈		غ	ჩ	Ռ
10000	[illegible]			[illegible]				M		萬	万				Μ	ⓐ				[illegible]	ჵ	O

so on, doubling each time. As a result, high-value binary numbers will be cumbersome and lengthy, simply because of the number of columns needed to place the bits. But for the computer, binary numbers are easy and rapid to read and record, thanks to the almost instantaneous speed of electrical pulses.

The binary system can be used equally effectively to express fractions. When written or printed, the number has a point to separate the whole number from the fraction, just as with a decimal number. So binary 101.101 will mean (reading from left to right): 4×1 (= 4) + 2×0 (= 0) + 1×1 (= 1) + (after the point) $\frac{1}{2} \times 1$ (= $\frac{1}{2}$) + $\frac{1}{4} \times 0$ (= 0) + $\frac{1}{8} \times 1$ (= $\frac{1}{8}$), otherwise $4 + 0 + 1 = 5$ (as the whole number) + $\frac{1}{2} + 0 + \frac{1}{8} = \frac{5}{8}$, otherwise $5\frac{5}{8}$, or, as a decimal fraction, 5.625.

2 NAMING THE NUMBERS

THE ORIGINS OF NUMBER WORDS

It is immediately obvious that the words for the numerals 1 to 10 are closely related in many European languages, whether the language is Romance (such as French), Germanic (German), Slavonic (Russian) or Celtic (Welsh). They are also directly related to the two European classical languages, Latin and Greek.

English	**French**	**German**	**Russian**	**Welsh**	**Latin**	**Greek**
one	un	ein	odin	un	unus	eîs
two	deux	zwei	dva	dau	duo	dúo
three	trois	drei	tri	tri	tres	treîs
four	quatre	vier	chetyre	pedwar	quattuor	téssares
five	cinq	fünf	pyat'	pump	quinque	pénte
six	six	sechs	shest'	chwech	sex	hex
seven	sept	sieben	sem'	saith	septem	heptá
eight	huit	acht	vosem'	wyth	octo	októ
nine	neuf	neun	devyat'	naw	novem	ennéa
ten	dix	zehn	desyat'	deg	decem	déka

At the same time, it may seem that some words in different languages for the same number are actually unrelated, such as Russian 'chetyre' and Welsh 'pedwar' for 'four'. But there is in fact a well-attested linguistic link between all such words, and any disparities will have resulted from the particular evolution of the given language, with changes in vowels and consonants, for example.

Historically, all the languages quoted here developed from a single original language. This was **Indo-European**, their parent language, which has been reconstructed and which produced many modern European and Indian languages, as well as classical Latin, Greek and Sanskrit. (For the Indo-European words for 1 to 10, see page 21.)

It is likely that the word for 'two' in many languages is additionally related to the word for 'you' ('thou') in the particular language. This reinforces the basic concept of 'I' (the speaker) as the grammatical 'first person' and of 'thou' (the person spoken to) as the 'second person', as well as tying in directly with the actual evolution of counting.

The relationship can be clearly appreciated as follows:

English	thou	two
French	tu	deux
German	du	zwei
Russian	ty	dva
Welsh	ti	dau
Latin	tu	duo
Greek	tu	dúo

Even though the vowels are not exactly similar for the two words, the consonants are, as *t* and *d* are directly related (*d* is simply a voiced *t*).

It will also be noticed that some consecutive numbers in a single language are unusually alike, such as Russian *sem'* and *vosem'*, or Latin *novem* and *decem*. This similarity probably arose through the linguistic 'attraction' of one number for its neighbour, which thus came to resemble it.

THE TEENS

On the whole, although by no means regularly, the numbers from 11 to 19 are mostly based on the word in a particular language for 'ten'.

In **English**, however, this 'ten' base does not appear until 13: thir*teen*, four*teen*, fif*teen*, six*teen*, seven*teen*, eigh*teen*, nine*teen* (the true 'teens', in fact). (For the detailed origin of these and the lower English numbers, including 11 and 12, see page 21.)

In **French**, the 'ten' base appears straightaway, at first simply as the final *-ze*, then in full from 17, where the order of the two numeral halves of the word actually becomes reversed: on*ze*, dou*ze*, trei*ze*, quator*ze*, quin*ze*, sei*ze*, *dix*-sept, *dix*-huit, *dix*-neuf. This alteration of formation at 17 (which also occurs in languages related to French, such as Italian and Spanish) probably arose from the original concept of 17 as a 'step number', beginning a new stage of counting. Up to 16, there was probably a '4-count', on the lines of 2 × 4 = 8, 3 × 4 = 12, and 4

× 4 = 16. On the other hand, it is possible that the alteration occurred to mark a kind of 'backward counting' from 20, reflecting the Latin, where 18, for example, is *duodeviginti*, that is, 'two from twenty'.

German numbers from 11 are like English, with the 'ten' base appearing only at 13: elf, zwölf, drei*zehn*, vier*zehn*, fünf*zehn*, sech*zehn*, sieb*zehn*, acht*zehn*, neun*zehn*.

The implication for numbers formed on base 'ten' is that the two numbers are added together, or joined, to form the higher number, so that English *fourteen* is 'four-ten' (or 'four-and-ten'), French *quatorze* is similar, although actually representing a development from Latin *quattuordecim*, 'four-ten', and German *vierzehn* is quite clearly 'four-ten', with neither of the basic numbers (4 and 10) altered in form.

With the **Russian** numbers from 11 to 19, the plan is similar, with however the actual word for 'on' (*na*) appearing between the lower number and 'ten'. This means that the Russian 'teens' are rather longer words than in the other languages considered here: odinna*dtsat'* ('one on ten'), dvena*dtsat'*, trina*dtsat'*, chetyrna*dtsat'*, pyatna*dtsat'*, shestna*dtsat'*, semna*dtsat'*, vosemna*dtsat'*, devyatna*dtsat'*.

Welsh numbers in the teens are more complex, and there are also alternative ways of expressing the same number. They run as follows, with the 'ten' element in italics, as for the numbers above:

11 un ar *ddeg* ('one on ten')
12 deu*ddeg* ('two-ten'), un *deg* dau ('one ten two')
13 tri ar *ddeg* ('three on ten'), un *deg* tri ('one ten three')
14 pedwar ar *ddeg*
15 pym*theg* ('five-ten')
16 un ar bym*theg* ('one on fifteen'), un *deg* chwech
17 dau ar bym*theg*, un *deg* saith
18 deunaw ('two nines'), un *deg* wyth
19 pedwar ar bym*theg*, un *deg* naw

Here it will be seen that 18 is a distinctive 'step number' and that 15 is used as a base for 16, 17 and 19.

The **Latin** numbers from 11 to 19 were directly responsible for producing the French teens, and the development can be easily traced. The 'ten' base is present down to 17: un*decim*, duo*decim*, tre*decim*, quattuor*decim*, quin*decim*, se*decim*, septen*decim*.

The regular words for 18 and 19, duodeviginti, undeviginti, translate respectively as 'two from twenty' and 'one from twenty', and so are a 'back count' from 20, the larger, round number attracting those near it. But there were also alternative words for 18 and 19, octo*decim* and noven*decim*, formed regularly like the earlier teens, and preserving the 'ten' base.

The equivalent **Greek** numbers are similar, but preserve the 'ten' link throughout. A small distinction is that the numbers 13 to 19 consist of the two lower numbers joined by the word for 'and' ('kai'): hen*déka*, do*déka*, treîs kai *déka*, téssares kai *déka*, pente kai *déka*, hex kai *déka*, heptá kai *déka*, októ kai *déka*, ennéa kai *déka*.

In modern Greek, however, 13 to 19 go 'ten-three', 'ten-four', etc: *deka*treîs, *deka*téssares, *deka*pénte, *dek*áxi, *dek*ephtá, *dek*okhtó, *dek*ennéa.

THE TENS

For the tens from 20 to 90, **English** uses a numeral form clearly related to the digits 'two', 'three', 'four', etc., with the final *-ty* related to 'ten', so that the numbers can be understood as 'two tens', 'three tens', etc: twen*ty*, thir*ty*, for*ty*, fif*ty*, six*ty*, seven*ty*, eigh*ty*, nine*ty*. (For more detailed comments on these, see below.)

As with the lower numbers, the **French** tens derive from the Latin, except for a marked change at 70, when a count based on base 20, not 10, begins, so that 70 is 'sixty-ten', 80 is 'four twenties' and 90 is 'four twenties-ten': vingt, trente, quarante, cinquante, soixante, soixante-dix, quatre-vingts, quatre-vingt-dix. Put technically, the French tens thus have a decimal gradation to 60, followed by a vigesimal gradation. The 20-count arose in France in the 11th century, replacing the earlier 10-count, which for 70, 80 and 90 had a form still found in French-speaking countries outside France today, such as Belgium: septante, huitante, nonante.

This 20-count was even used above 100, so that 120, for example, was six-vingt ('six twenties') and 140 was sept-vingt ('seven twenties'). The vigesimal count went up as high as 360 (dix-huit-vingt), and although now regularly represented only by 80 (quatre-vingts) is still found in a few historic titles, such as les Quinze-Vingts ('the fifteen twenties'), the hospital founded for 300 blind people in Paris by Louis IX in the 13th century. Its name remains today for the eye hospital on the site.

The **German** tens are straightforward, like the English, with the final *-zig* corresponding to the English *-ty*: zwan*zig*, drei*ßig*, vier*zig*, fünf*zig*, sech*zig*, sieb*zig*, acht*zig*, neun*zig*.

With one striking exception (40), the **Russian** tens proceed much as in English, meaning 'two tens', 'three tens', etc: dvadtsat', tridtsat', sorok, pyat'desyat, shest'desyat, sem'desyat, vosem'desyat, devyanosto.

The distinctive word for 40 indicates a 'fullness' or 'completion' as a rounded total, and can be equated with the many proverbial allusions to 40 still found in English (such as 'forty days and forty nights'). The actual derivation of Russian *sorok* is disputed. Some have attempted to take it from the Greek word for 40, *tessarákonta*. But this is unlikely on both linguistic and historic grounds. Others have derived it from Old Norse *serkr*, 'coat', so that the word refers to the quantity of sable skins required to make such a coat. But the word may actually have developed from Turkish, where the word for 40 is *kirk*.

Russian 90 is also slightly different. The word (*devyanosto*) is clearly based on *devyat'* (9) with the latter half borrowed from the next big number up, *sto* (100). At the same time, it may have assumed a different form due to the traditional Russian custom of reckoning by nines, traces of which still remain in folk tales. (The expression *za tridevyat' zemel'*, for example, means literally 'beyond three-nines lands', approximating to English 'at the other end of the world'.)

The **Welsh** word for 20, *ugain*, is related to Latin *viginti*, just as the French *vingt* more obviously is. As in the teens, the Welsh tens have alternative names:

20 ugain
30 deg ar hugain ('ten on twenty'), tri deg ('three tens')
40 deugain ('two twenties')
50 hanner cant ('half-hundred'), deg ar deugain
60 trigain ('three twenties'), chwe deg ('six tens')
70 deg a thrigain ('ten and sixty'), saith deg ('seven tens')
80 pedwar ugain ('four twenties'), wyth deg ('eight tens')
90 deg a phedwar ugain ('ten and eighty'), naw deg

It will be seen that there is not only a vigesimal (20-based) count here, but that 50 is thought of as a round number in terms of half a hundred.

The vigesimal principle is a regular feature of other Celtic languages beside Welsh, such as Irish and Scottish Gaelic.

Latin tens are regular, with the *-ginti* of 20 and the *-ginta* of higher tens representing 'ten' in the same way that English *-ty* and German *-zig* do: vi*ginti*, tri*ginta*, quadra*ginta*, quinqua*ginta*, sexa*ginta*, septua*ginta*, octo*ginta*, nona*ginta*.

The *-ginta* is in turn related to the *-konta* that ends the **Greek** tens, although in altered form in 20: eí*kosi*, triá*konta*, tessará*konta*, pent*ḗkonta*, hex*ḗkonta*, hebdom*ḗkonta*, ogdo*ḗkonta*, emen*ḗkonta* (70, 80 and 90 are based on ordinals).

For intermediate numbers, such as 21, 32, 43, etc., all these languages use a straightforward combination of the lower number with the higher, whether preceding or following it, and in some cases joined by the word for 'and'. For 25, for example, French has *vingt-cinq*, German *fünfundzwanzig*, Russian *dvadtsat' pyat'*, Welsh *dau ddeg pump*, Latin *quinque et viginti*, Greek *pénte kai eíkosi*.

THE HUNDREDS

Although in some cases not obviously so, the words for 100 in all six languages are related: English *hundred*, French *cent*, German *hundert*, Russian *sto*, Welsh *cant*, Latin *centum*, Greek *hekatón*.

All six languages, too, form the hundreds from 200 to 900 in a fairly regular manner, such as French *deux cents*, *trois cents*, etc., German

POCKETS OF PEAS

The ancient system of reckoning in groups of numbers, like the tens scored through on the primitive tally stick, has been put to practical use in more recent times. When the famous German guidebook publisher, Karl Baedeker, was preparing his volume on Italy, he counted the many steps up to Milan Cathedral by removing one pea from his waistcoat pocket every 20 steps and putting it into his trouser pocket.

UNIVERSAL SEVEN

The word for 'seven' is related not only in Indo-European languages but in the Semitic languages, which include Arabic and Hebrew, and the root element for the number seems even to tally in other languages outside these groups. In Indo-European languages, 'seven' is found as Latin *septem*, Irish *seacht*, Gothic *seabun*, Greek *heptá* (originally *septá*), Albanian *shtátë*, Slavonic *sedm'* (modern Russian *sem'*), Lithuanian *septyni*, Sanskrit *saptan*, Iranian *haft*, Hittite *śipta* and Tocharian *śpät*, to say nothing of more familiar languages such as French *sept* and German *sieben*. In Semitic languages 'seven' is *sab'* in Arabic, *śeva'* in Aramaic, *sebgha* in Maltese, *sebat* in Amharic, *śav'o* in Syriac (a dialect of Aramaic) and *sıbu* in Akkadian, once spoken in Mesopotamia. Then in Finnish it is *seitsemän*, in Hungarian *hét*, in Egyptian *sefen̂*, in Coptic *śaśf*, in Tuareg *sa*, in Swahili *saba*, in Khmer *satta* and in Malay *sapta*.

zweihundert, *dreihundert*, etc., Russian *dvesti*, *trista*, etc. (the form of *sto* changes, and from 500 is *sot*), Welsh *dau gant*, *tri cant*, etc., Latin *ducenti*, *trecenti*, etc., Greek *diakósioi*, *triakósioi*, etc.

Intermediate numbers involve a reasonably straightforward compilation of the three figures (units, tens, hundreds), such as: French *quatre cent trente-six* (436), German *sechshundertsiebenundachtzig* (687, written as a single word), Russian *pyat'sot sorok odin* (541), Welsh *dau gant a naw* (209, literally 'two hundred and nine'), Latin *octingenti viginti sex* (826), Greek *pentakósioi téssares kai hexḗkonta* (564).

THE THOUSANDS AND BEYOND

Although the words for 'hundred' in the six languages are all related, those for 'thousand' are not. The English thousand and German *tausend* are akin, however, as are French *mille*, Welsh *mil*, Latin *mille* and possibly Greek *khílioi*. Russian *tysyacha* is out on its own, and may originally have meant something like 'big hundred'.

Intermediate numbers between 1000 and 1 million simply involve stating the number of thousands followed by the hundreds in the usual way.

'Million' is *million* in French, *Million* in German, *million* in Russian, *miliwn* in Welsh, but *deciens centena milia* ('ten times a hundred thousand') in Latin, which had no single word for 'million'. Nor did Greek, although it did have *mȳriás* for 'ten thousand'.

The word 'million', itself found in many modern European languages, relates directly to Italian 'millione', where the *-one* suffix is an augmentative, meaning 'big', so that the literal meaning is 'big thousand'.

Numbers above 'million' are essentially modern words, adopted to express scientific quantities. Like many scientific terms themselves they are based on Latin, so that 'billion' is formed from *bi-* ('two') + million; 'trillion' consists of *tri-* ('three') + million; and so on (quadrillion, quintillion, etc.).

HOW MANY IN A MILLION?

'Million' means 1000 times 1000, in other words 1,000,000, so is clearcut. 'Billion', however, can mean either '1 million times 1 million' (1 with 12 zeros) or '1000 times 1 million' (1 with 9 zeros). The former is the common meaning in Britain, France and other continental countries, while the latter is frequently found in the USA. It also

GREEK NUMERICAL PREFIXES

The Système International d'Unités (SI) has been internationally adopted as a system of units for the measurement of all physical quantities. It uses Greek-derived prefixes to indicate multiples and submultiples in order to avoid a multiplicity of zeros when expressing very large or very small metric amounts. The prefixes, together with the standard abbreviations and factors of the numbers, are given below.

Prefix	*Abbrev.*	*Factor*
exa-	E	million million million (10^{18})
peta-	P	thousand million million (10^{15})
tera-	T	million million (10^{12})
giga-	G	thousand million (10^{9})
mega-	M	million (10^{6})
kilo-	k	thousand (10^{3})
*hecto-	h	hundred (10^{2})
*deca-	da	ten (10)
*deci-	d	tenth (10^{-1})
*centi-	c	hundredth (10^{-2})
milli-	m	thousandth (10^{-3})
micro-	μ	millionth (10^{-6})
nano-	n	thousand millionth (10^{-9})
pico-	p	million millionth (10^{-12})
femto-	f	thousand million millionth (10^{-15})
atto-	a	million million millionth (10^{-18})

* used only where kilo- or milli- would be impractical

occurs in the UK, however, which can make for confusion. The higher number is of practical use only to those scientists, such as astronomers, who deal in very large numbers. The lower figure obviously has a more common scientific application, and is a useful word for the quantity.

There is also a word 'milliard' to mean '1000 times 1 million' but it is rarely used, which is a pity, as it could fill this gap very well, leaving 'million' for the higher number.

The compound million words have a similar double sense, so that 'trillion' can mean 'million million million' (1 with 18 zeros), as usually in Britain, or 'million million' (1 with 12 zeros) in the USA.

Higher compounds have the following forms and values:

	Number of zeros in UK	*Number of zeros in USA*
quadrillion	24	15
quintillion	30	18
sextillion	36	21
septillion	42	24
octillion	48	27
nonillion	54	30
decillion	60	33
undecillion	66	36
duodecillion	72	39
tredecillion	78	42
quattuordecillion	84	45
quindecillion	90	48
sexdecillion	96	51
septendecillion	102	54
octodecillion	108	57
novemdecillion	114	60
vigintillion	120	63
centillion	600	303

Centillion is thus the highest lexicographically recognized named number, and is recorded by the *Oxford English Dictionary* as occurring (in adjectival form, 'centillionth') in 1852.

THE ORIGINS OF ENGLISH CARDINAL NUMBERS

Almost all the words for the English cardinals derive from Old English (Anglo-Saxon), the form of the English language that was spoken before about AD 1100. Ultimately, such words in turn can be traced back to an Indo-European source, as

can the numerals in many other European languages.

The pedigree of the individual words can be summarized as follows (OE means 'Old English'; IE means 'Indo-European'; an asterisk (*) means a reconstructed word):

one OE *ān* ← IE **oinos*
two OE *twā* ← IE **dwō*
three OE *thrī* ← IE **trejes*
four OE *fēower* ← IE **qetwōr*
five OE *fīf* ← IE **pempe*
six OE *siex* ← IE **sweks*
seven OE *seofon* ← IE **septḿ̥*
eight OE *ehta* ← IE **oktō*
nine OE *nigon* ← IE **newn* (related to IE **newjos*, 'new')
ten OE *tēn* ← IE **dek̂m̥*
eleven OE *endleofon*, literally 'one left', i.e. 10 + 1
twelve OE *twelf*, literally 'two left', i.e. 10 + 2
thirteen OE *threōtīene*, i.e. 3 + 10
fourteen OE *fēowertīene*, 4 + 10
fifteen OE *fīftēne*, 5 + 10
sixteen OE *siextīene*, 6 + 10
seventeen OE *seofontīene*, 7 + 10
eighteen OE *eahtatēne*, 8 + 10

THE GOOGOL

A googol is a very high number: 10 to the power of 100, or 1 followed by 100 zeros. It was written in the 1930s on the blackboard of a New York kindergarten school thus (as it doubtless has been before and since):
100000000000000000000000000000000000000000000000000
00000000000000000000000000000000000000000000000000.
The American mathematician Edward Kasner (1898–1955) asked his nine-year-old nephew, Milton Sirotta, what he would call this number. The boy came up with the word 'googol', no doubt fancifully based on 'goo-goo' or some other childish word or name. The same youngster, it seems, also invented the term 'googolplex' for an even higher number: 1 followed by a googol of zeros. Mathematicians were glad to have a name for one of the important high numbers in which they increasingly had to deal, although the term has never been formally adopted.

nineteen OE *nigontȳne*, 9 + 10
twenty OE *twentig*, literally 2 10s
thirty OE *thrītig*, 3 10s
forty OE *fēowertig*, 4 10s
fifty OE *fīftig*, 5 10s
sixty OE *siextig*, 6 10s
seventy OE *seofontig*, 7 10s
eighty OE *eahtatig*, 8 10s
ninety OE *nigontig*, 9 10s
hundred *hundred* ← IE **kṃtóm* (based on the IE for '10') + Germanic base **rath*, 'number', so really 'big number 10'
thousand OE *thūsend* (but no earlier IE word); first part of word perhaps related to '*tum*id' and second to 'hund*red*', so perhaps with overall sense 'swollen hundred'
million Old French *million* ← Old Italian *mille* (from Latin *mille*, 'thousand') + suffix *-one*, 'big', so 'big thousand'
billion French *billion*, with Latin *bi-*, 'two', substituted for the *mi-* of *million*
The higher multiples (*trillion* etc.) are formed similarly.

THE ORIGINS OF ENGLISH ORDINAL NUMBERS

The English ordinal numbers have a similar origin (except 'second'), with the number itself clearly related to the corresponding cardinal:

first OE *fyrst* ← IE root element **pr*, found also in modern prime, prince, prior, etc.
second Latin *secundus*, 'following', 'favourable', itself from *sequi*, 'to follow' (as in modern English *sequence*); original word for 'second' was 'other' (OE *ōther*); compare modern 'one or the other', 'one another', etc.)
third OE *third*, a variant of *thridda* ← IE **tritjos*
fourth OE *feowertha*
fifth OE *fifta*
sixth OE *siexta*

Higher ordinals are similar, with the final *-th* of the units and teens, and the *-eth* of the tens representing OE *otha* or *othe*, which has its parallel in other languages including Latin (*-tus*, as in quar*tus*) and Greek (*-tos*, as in tetra*tos*, 'fourth').

ENGLISH NUMERAL COGNATES

Many standard words in English are cognates (linguistic relatives) of the basic cardinal numerals. Among them are:

one a, alone, an, anon, atone, lone, none, nonce, once, only.
two between, twain, twice, twig (as a forking branch), twilight (as half day, half night), twill (as woven from two threads), twin, twine (as having two strands), twist (as turning twice).
three drill (as woven from three threads), thrice, trellis (as having three strands), plus many words with a 'threefold' sense beginning *ter-* or *tri-*, which usually derive from Latin or Greek (see **Classical Cognates**).
four farthing (as a quarter of a penny), firkin (as a quarter of a barrel), plus many Latin or Greek words beginning *quad-* or *tetra-*.
five fin (US slang, '$5 bill') plus Latin and Greek words beginning *quin-* or *penta-*.
six semester (as properly a six-month term: Latin *sex*, 'six', and *mensis*, 'month'); siesta (as a rest at the sixth hour of the day, i.e. noon), plus some Latin or Greek words beginning *sex-* or *hexa*.
seven sennight (old word for 'week', as has seven nights, compare French *semaine*, 'week'), plus Latin and Greek words beginning *sept-* or *hepta-*.
eight Latin or Greek words beginning *oct-*.
nine noon (as originally the ninth hour of the day, i.e. 3.00, but 'attracted' to 12.00 as forming a more natural break for a meal and a rest) plus Latin words beginning *nov-*.

ten dean (ultimately Latin *decanus*, as one in charge of ten men), dime (ultimately Latin *decima*, 'tithe', as one tenth of a dollar), tithe (literally 'tenth', as originally the tenth part of produce paid to the Church) plus Latin and Greek words beginning *dec-*.
fourteen fortnight (as having 14 nights).

Other cognates are formed much more obviously by adding a suffix to the English cardinal, such as the series twoer, three-er, fourer, fiver, sixer, etc. (thing or person associated with the respective number), twofold, threefold, fourfold, etc. (multiplied the stated number of times), twosome, threesome, foursome (comprising two, three, four, people or things).

OTHER ORIGINS

The origins of the words for some numbers that do not form part of the regular sequence are as follows:

half (½) Old English *half*, ultimate origin obscure
dozen (12) Old French *dozeine* ← Latin *duodecim*, 'twelve'
score (20) notch on tally stick
pony (£25) perhaps as small sum, as pony is small horse
gross (144) French *grosse*, 'big', for goods in bulk
monkey (£500) perhaps based on 'pony', but 'cleverer'
chiliad (1000) Greek *khilioi*, 'thousand' + *-ad*, 'group'
myriad (10,000) Greek *myrioi*, '10,000' + *-ad*, 'group'
lakh (100,000 rupees) Hindi *lākh*

TREBLE OR TRIPLE?

The two words meaning 'multiple of three' vie with each other in a range of expressions. We have *treble* chance but *triple* jump, *treble* clef but *triple* time in music, *treble* bob in bellringing, but *Triple* Crown in sport.

And what is 'threefold' about the musical *treble* clef? It may originally have served for the melody sung by the *third* voice above the tenor and alto, say some. But *treble* means 'threefold', not 'third', and this explanation, the one usually given, is not very satisfactory. The precise origin of the treble clef remains a mystery.

SQUARE WORDS

The English word 'square' derives from Latin *quadra*, which itself means 'square' and derives from *quattuor*, 'four'. As *four* is itself related to Latin *quattuor*, both words are therefore linguistically linked to 'square'.

But 'square' has many other relations. Among the closest are *squad* and *squadron*. A 'squad' was originally a group of 12 soldiers who drilled in a square formation, and a squadron (from Italian *squadrone*) was simply a larger such drill formation.

Still in the military field, and also related to 'square', is a *cadre*. The word is French for 'frame', and came to denote the 'framework' of officers and under-officers who formed the permanent establishment of a regiment.

Yet another 'square' relation is the *quarry*, where square blocks of stone are cut from the rock. The word came into English from Latin via Old French *quarriere* (modern French *carrière*).

crore (10 million rupees) Hindi *karōr*
googol (10^{100})
googolplex (10^{googol})

0, NOTHING, NOUGHT, NIL, NIX, ZERO, ZILCH . . .

The words for 0 vary in origin. 'Nought' is really 'no aught', or 'not anything', as in the slightly different spelling 'naught'. To 'come to naught' is to end as nothing. 'Nil' is a contraction of Latin *nihil*, itself from *ni hilum*, 'no small thing'. 'Nix' comes from German *nichts*, 'nothing'.

'Zero' and 'cipher' are words of related origin, both ultimately originating in Arabic *sifr*, 'cipher' (compare French *chiffre*, 'figure', and the English 'decipher'). The basic sense of the Arabic word itself was 'empty'.

'Zilch' is a slang word of uncertain origin, but it was probably influenced by the *z-* of 'zero' and the word 'million'. ('Zillions' is a similar slang term for the opposite: 'many millions'.)

CLASSICAL COGNATES

Many scientific and specialist words are based on Latin or Greek numeral words, especially those for 1 to 10. For 'one' words, Latin *uni-* and Greek *mono-* are the usual elements used, and for 'two' words Latin *bi-* and Greek *di-*. Some words combine a Latin numeral with a Greek word, and vice versa.

A list of some of the most familiar words for each numeral follows, with a literal translation in brackets. The list also includes some words that are popularly, but wrongly, associated with the numeral (such as 'bikini'). Where the two languages combine in a single word, the second part is prefixed accordingly by [L] or [G], for 'Latin' or 'Greek'. The modern meanings of the words can be checked, as necessary, in a standard dictionary.

one LATIN: *unanimous* (one mind), *unicorn* (one horn), *unicycle* (one [G] wheel), *uniform* (one form), *unilateral* (one-sided), *union* (being one), *unique* (one and only), *unison* (one sound), *unit* (one digit), *universe* (turned into one), *university* ('universal' or complete body of scholars and students).
GREEK: *monarch* (one rule), *monastery* (living alone), *monaural* (one [L] ear), *monk* (living alone), *monochrome* (one colour), *monocle* (one [L] eye), *monogamy* (one marriage), *monoglot* (one tongue), *monogram* (one writing), *monograph* (one writing), *monokini* (one-piece bikini, see latter word under *two* below), *monolith* (one stone), *monologue* (one word), *monopoly* (one seller), *monotonous* (one tone).

two LATIN: *biathlon* (two [G] contests: see panel below), *biceps* (two-headed), *bicycle* (two [G] wheels), *biennial* (two years), *bigamy* (two [G] marriages), *bikini* (two-piece swimsuit, wrongly associated with *bi-*, as actually named after Pacific island of Bikini), *bilateral* (two-sided), *binary* (two together), *binoculars* (two eyes), *biped* (two feet), *biplane* (two planes), *bisect* (cut in two).
GREEK: *dichotomy* (cut in two), *dilemma* (two assumptions), *dimity* (two threads), *diphthong* (two voices), *diploma* (folded in two).

three LATIN: *triangle* (three angles), *triceps* (three-headed), *tricolour* (three colours), *tricorn* (three horns), *tricycle* (three [G] wheels), *triennial* (three years), *trikini* (three part-bikini, see latter word under 'two', above), *trimaran* (three-hulled catamaran), *trinity* (set of three), *triple* (threefold), *triplet* (set of three), *trishaw* (tricycle rickshaw), *triumvirate* (three-man rule), *trivet* (three feet), *trivial* ('of three ways', i.e. crossroads talk).
GREEK: *triad* (set of three), *triarchy* (government by three persons), *trigonometry* (three angle measure), *trigraph* (three writings), *trilith* (three stones), *trilogy* (three words), *tripod* (three feet), *tripos* (three feet, from three-legged stool on which disputer sat at degree ceremony), *triptych* (three-folded), *triskelion* (three-legged, as symbol of Isle of Man).

four LATIN: *quadrangle* (four angles), *quadrant* (four-part), *quadraphonic* (four [G] sounds), *quadrilateral* (four-sided), *quadriplegic* (four blows, as paralysed in all four limbs), *quadroon* (fourth, as a quarter Negro blood), *quadruped* (four feet), *quadruplet* (set of four), *quart* (fourth of a gallon), *quarter* (fourth part).
GREEK: *tetrameter* (four measures), *tetrarch* (four ruler, i.e. over one of four divisions of a country), *trapezium* (four feet, i.e. like a table).

five LATIN: *quincunx* (five-twelfths, as five-twelfths of an *as* [the Roman pound of 12 ounces] was denoted by five dashes arranged as in a *quincunx*), *quintain* (fifth in rank, from Latin *quintana*, street in Roman camp separating fifth maniple [subdivision of legion] from sixth, where military exercises were carried out), *quintessence* (fifth essence, i.e. after four elements of earth, fire, air, water), *quintuplet* (set of five).

OLYMPIC GAMES

Many Olympic sports comprise a multiple event, denoted by a Greek word consisting of a numeral followed by *athlon*, 'contest' (compare modern 'athlete', 'athletics'). In Ancient Greece, the best known was the *pentathlon* ('five contests'), held on the second day of the five-day games. This comprised running, jumping, throwing the discus, throwing the javelin and wrestling. The Greek terms for the five events have been preserved, appropriately, in a single pentameter ('five-feet') line of verse: *hálma, podókeian, dískon, ákonta, pálen* (Simonides, *Lyra Graeca*).

In the Modern Games, the contests are as follows:
biathlon ('two contests'): cross-country skiing and rifle-shooting (in Winter Olympics);
pentathlon ('five contests'): (1) (for women) 100-metre hurdles, shot put, high jump, long jump and 200-metre sprint; (2) (modern pentathlon) 800-metre equestrian show-jumping, épée fencing, 300-metre freestyle swimming race, target shooting at 25 metres and 4000-metre cross-country run;
heptathlon ('seven contests'): (for women) 100-metre hurdles, high jump, shot put, 200-metre sprint, long jump, javelin and 800-metre race;
decathlon ('ten contests'): 100-metre race, long jump, shot put, high jump, 400-metre race, 110-metre high hurdles, discus, pole vault, javelin and 1500-metre race.

> **THE LENGTH OF A SAROS**
>
> As an extension of their hexagesimal (60-based) system of reckoning, the Babylonians used the word *saros* (partly of Assyrian origin) to denote the number 3600. The term came to be used to express this number of years, over which a series of lunar and solar eclipses were calculated to recur. The word is still used by astronomers today, but it now denotes a period of 6585 *days* and 8 hours, after which the relative positions of the Sun and Moon recur. The different value given to *saros* is due to the 10th-century Greek lexicographer Suidas, who because of a misunderstanding or misinterpretation took the word to mean a period of 18½ years, not 3600 as originally.

GREEK: *pentagon* (five angles), *pentameter* (five measures), *Pentateuch* (five books, as first five of Bible), *pentathlon* (five contests).

six LATIN: *sextant* (sixth part, as measures one-sixth of a circle).
GREEK: *hexameter* (six measures).

seven LATIN: *September* (seventh month, in original calendar, see p. 34), *septentrional* (seven plough-oxen, i.e. seven stars of Great Bear, which lie in northern sky).
GREEK: *heptameter* (seven measures), *heptathlon* (seven contests).

eight LATIN: *octane* (eight alk*ane*, for eight carbon molecules in the liquid, whose formula is C_8H_{18}), *octave* (eighth, i.e. eighth note of scale), *October* (eighth month, in original calendar).
GREEK: *octagon* (eight angles), *octopus* (eight feet).

nine LATIN: *November* (ninth month, in original calendar), *novena* (nine, i.e. devotion lasting nine days).
GREEK: *nonagon* (nine angles).

ten LATIN: *December* (tenth month, in original calendar), *decimal* (tenth), *decimate* (tenth, i.e. kill one in ten).
GREEK: *decade* (group of ten), *Decalogue* (ten words, i.e. the Ten Commandments), *decathlon* (ten contests).

The classical tens also produce some familiar words, whether or not through an intermediary language, such as French. They include:

forty LATIN: *Quadragesima* (fortieth), *quarantine* (forty, as original length of period in days).

fifty LATIN: *Quinquagesima* (fiftieth).
GREEK: *Pentecost* (fiftieth, as observed on 'fiftieth day' [actually 49th] after Easter).

sixty LATIN: *Sexagesima* (sixtieth).

seventy LATIN: *Septuagesima* (seventieth), *Septuagint* (seventy, as 'seventy translators' of Old Testament into Greek).

hundred LATIN: *cent, per cent* (hundred), *centenary* (containing a hundred), *centigrade* (hundred degrees), *centipede* (hundred feet), *centurion* (commander of a hundred men, i.e. a century), *century* (group of a hundred).

CLASSICAL-BASED NUMERICAL SEQUENCES

Just as English has numerical sequences such as once, twice, thrice or twosome, threesome, foursome, so there are several sequences based on Greek or Latin, in some instances developing through a modern language such as Italian or French, and either having an added word or element (like the English *-some*) or not.

Here are some of the most familiar, with an interpretation of the added word or element, where applicable, and an indication of the sequence's area of usage. Some individual examples have already occurred in the listing above.

Members of a sequence that do not linguistically conform (such as solo) are included in brackets. The value of each member of a sequence is additionally given after it.

WITH ADDED WORD OR ELEMENT:

-ad, 'group' (Greek), used in numerology to express a number: monad (1), dyad (2), tryad (3), tetrad (4), pentad (5), hexad (6), heptad (7), octad (8), ennead (9), decad (10)

> **THE OLD HUNDRED**
>
> In English history, the 'hundred' was a former subdivision of a county or shire, and had its own court. But why was it so named? Opinions differ, but it seems likely that the hundred was assessed, for purposes of taxation, at a value of 100 hides, a hide being the amount of land that would support a single household. In practice, however, the hundreds were rated at both higher and lower figures than this. The word survives in the expression 'to apply for the Chiltern Hundreds', used of a Member of Parliament who wishes to resign from the House of Commons. The post of stewardship of the Chiltern Hundreds is now a nominal one, although originally the office would have involved the suppression of robbers who frequented the wooded Chiltern Hills, in Buckinghamshire.

0—NOTHING OR SOMETHING?

What kind of a number is 0? Is it a digit or isn't it? 1, 2, 3, 4, 5, 6, 7, 8 and 9 are all digits that can be instantly understood. But how about 0? It is usually regarded as 'nothing'. But sometimes it is 'nothing' and at other times it is definitely 'something'. 3 + 0 = 3, for example, as does 3 − 0. Here, 0 is 'nothing' added or subtracted. But when 0 follows another number, as in 30, it suddenly multiplies that number by 10, so it is clearly 'something'. And in a number such as 40,000, the string of 'nothings' multiplies the 4 ten thousand times. No wonder that a French writer of the 15th century called 0 *un chiffre donnant umbre et encombre*, 'a figure causing confusion and difficulty'.

-centenary, from Latin *centum*, 'hundred', used to express a hundredth anniversary: bicentenary (200), tercentenary (300), quatercentenary (400), quincentenary (500), sexcentenary (600), septcentenary (700), octocentenary (800), novocentenary (900)

-et, 'group' (Italian, from Latin), used in music to denote the number of players or singers: (solo) (1), duet, (duo) (2), (trio) (3), quartet (4), quintet (5), sextet (6), septet (7), octet (8), nonet (9)

-genarian, from Latin *-ginta*, 'tens', used to express a person's age in the tens: quadragenarian (40s), quinquegenarian (50s), sexagenarian (60s), septuagenarian (70s), octogenarian (80s), nonagenarian (90s), (centenarian) (100+)

-gon, 'angle', from Greek *gonia*, 'angle', used to denote a geometrical figure with a given number of angles and sides: (triangle) (3), tetragon, (rectangle, quadrilateral, square) (4), pentagon (5), hexagon, (cube) (6), heptagon (7), octagon (8), nonagon (9), decagon (10), hendecagon (11), dodecagon (12)

-ple, '-fold', from Latin *plicare*, 'to fold', used to denote a group or multiple: (single) (1), (double) (2), triple, (treble) (3), quadruple (4), quintuple (5), sextuple (6), septuple (7), octuple (8), nonuple (9), decuple (10), centuple (100)

-plet, 'multiple', from Latin as above, but in form based on 'doub*let*', used to denote a multiple, but especially a multiple birth: (twins) (2), triplets (3), quadruplets (quads) (4), quintuplets (quins) (5), sextuplets (6), septuplets (7), octuplets (8), nonuplets (9), decuplets (10)

-plicate, '-fold', from Latin as above, used to denote the number of identical copies: duplicate (2), triplicate (3), quadruplicate (4), quintuplicate (5), sextuplicate (6), septuplicate (7), octuplicate (8), nonuplicate (9), decuplicate (10), centuplicate (100)

-reme, from Latin *ramus*, 'oar', used to denote a Roman galley with a given number of banks of oars: unireme (1), bireme (2), trireme (3), quadrireme (4), quinquereme (5).

Other sequences are usually more specialized and with a restricted numerical run. They include:

-cento, Italian 'hundred' (from Latin *centum*), used to denote a particular century in literature or art: trecento (14th, which began in 1301), quattrocento (15th, which began in 1401), cinquecento (16th), seicento (17th), settecento (18th)

-ceps, from Latin *caput*, 'head', used to denote a muscle with a particular number of 'heads' or origins: biceps (two, at front of upper arm), triceps (three, at back of upper arm), quadriceps (four, at front of thigh)

-cycle, from Greek *kyklos*, 'wheel', used to denote a vehicle (usually pedalled) with a particular number of wheels: monocycle, (unicycle) (1), bicycle (2), tricycle (3), quadricycle (4)

-gamy, from Greek *gamos*, 'marriage', used to denote a particular number of marriages: monogamy (1), bigamy (2), trigamy (3)

-plane, from Latin *planus*, 'level', but formed on 'aero*plane*', used to denote an aircraft with a given number of pairs of wings (planes): monoplane (1), biplane (2), triplane (3), quadriplane (4)

-plegia, from Greek *plege*, 'strike', 'blow', used to denote paralysis in the stated number of parts of the body (usually, limbs): monoplegia (1), diplegia (2), triplegia (3), quadriplegia, tetraplegia (4); also hemiplegia, from Greek *hemi-*, 'half' to denote paralysis in one (lateral) half of the body.

(For the sports series biathlon, pentathlon, etc., see box, p. 24.)

WITHOUT ADDED WORD OR ELEMENT:

1 used to denote a simple order, corresponding to English *first*, *second*, *third*, from Latin: *primary* (1st), *secondary* (2nd), *tertiary* (3rd), *quaternary* (4th);

THE ORIGINAL NUMBER

Where did the word 'number' itself originate?

It came into English from French *nombre*, which in turn came from Latin *numerus*. The Latin word is related to Greek *nemo*, 'I distribute', and *nomos*, 'law'. A near relation of the English word is *nimble*, which basically has the sense of 'taking', as in modern German *nehmen*, 'to take'.

'Number' is a similar word in many languages, including French *nombre*, German *Nummer*, Italian *numero*, Spanish *número* and Russian *nomer*.

The abbreviation *No.* comes from Latin *numero*, 'in number'.

2 used to denote certain of the seven 'Day Hours' of the monastic divine office of the Christian church, from Latin: (Lauds), *Prime* (1st hour, i.e. 6 a.m.), *Terce* (3rd hour, or 9 a.m.), *Sext* (6th hour, or 12 noon), *None* (9th hour, or 3 p.m.), (Vespers), (Compline);

3 used to denote the different types of parrying position in fencing, from French: *prime* (1st), *seconde* (2nd), *tierce* (3rd), *quarte* (4th), *quinte* (5th), *sixte* (6th), *septime* (7th), *octave* (8th);

4 used to denote paper sizes, especially for printed sheets, from Latin: *quarto* (folded into 4), *octavo* (into 8), *sextodecimo* (sixteenmo) (into 16), (thirty-twomo) (into 32), each folded sheet giving twice that number of pages;

5 used to denote a verse or stanza with a particular number of lines, from French: (*terzina*) (3), *quatrain* (4), *quintain* (5), *sexain* (*sixain*, *sextain*) (6);

6 used to denote the value of a playing card or die (dice): (*ace*) (1), *deuce* (2), *trey* (3), *quatre* (4), *cinque* (5), *sice* (6).

For the series of religious festivals, *Quadragesima*, *Quinquagesima*, etc., see box, p. 52.)

WORDS WITH HIDDEN NUMBER MEANINGS

Most of the words instanced above have obvious numerical origins, whether English or in a classical or modern language. But there are several less obvious, such as *diploma*, *trivial*, *tithe*. Here are some more of this type.

1 *ace*, from Latin *as*, 'unity', itself used as the name of a Roman coin.

2 *balance*, from Latin *bilanx*, literally 'double scale'; *biscuit*, from French *bis cuit*, 'cooked twice'; *combine*, from Latin *com-* 'with' and *bini*, 'two together'; *dubious*, from Latin *duo*, 'two', as hesitating between two alternatives, so therefore also *doubt*.

3 *sitar*, from Hindi, 'three-stringed', as this is the basic number of strings the Indian instrument has (although its actual number varies); *testament*, from Latin *testis*, 'witness', itself related to *tres*, 'three' and *stare*, 'to stand', as the witness stands by as a third party in a legal process; *travel*, from French *travail*, 'work', 'labour', from Latin *trepalium*, 'instrument of torture', from *tres*, 'three' and *palus*, 'stake'.

4 *charpoy*, from Hindi *chārpāī*, from Persian *chāhār-pāī*, 'four feet', as this is the number the Indian bedstead has; *quire*, from Old French *quaier* (modern French *cahier*, 'exercise-book'), from Latin *quaterni*, 'set of four', as originally four sheets of paper folded to make eight pages (but now the term for 24 sheets, or one-twentieth of a ream).

5 *punch*, from Hindi *pā̃c*, 'five', as this is the traditional number of the five original ingredients of the drink: spirit, water, sugar, lemon juice, spice (although some query this etymology).

6 *samite*, ultimately from Greek *hexa-*, 'six' and *mitos*, 'thread', as this is the number of threads

IMPLICIT NUMBERS

Some words *suggest* a particular number, or imply one, without actually stating it or containing a linguistic element that denotes it.

Such words are commonest for the low numbers, and include the following:

suggesting **1**: ace, hermit, individual, single, solo
suggesting **2**: both, brace, couple, geminate (to double), hermaphrodite (having both male and female sexual characteristics), love-seat (armchair for 2), pair, polarity (state of having 2 opposite poles), superpower (USA or USSR), tandem, tête-à-tête
suggesting **3**: cube (power of 3), deltoid (shaped like the Greek letter delta, so triangular), gooseberry (unwanted third person with two lovers), hat-trick (dismissing three batsmen with three successive balls in cricket), leash (three animals), shamrock (plant with three leaves)
suggesting **4**: cross, crossroads, mandala (Buddhist symbol of circle enclosing a square with a god on each of four sides)
suggesting **5**: hand (as group of five bananas, with five 'fingers'), lustrum (Roman period of five years).

MEANINGFUL NUMBERS

As well as its basic sense of 'figure', 'numeral', 'quantity', *number* is a word that has acquired various other meanings over the years.

Here are some, as given by the *Oxford English Dictionary*:

1 An issue of a book or periodical, as a 'recent number of the magazine';
2 A poem, song or item in a musical entertainment, as a 'novel number' or a 'witty number';
3 A person or thing, especially a woman or an article of clothing, as 'she's a lovely number' or a fashion designer's 'autumn number';
4 A job or piece or work, as 'a cushy number'.

woven or interwoven (in different colours) in the silk fabric.

7 *hebdomadal*, from Greek *hepta*, 'seven', so that the term means 'weekly', hence the Hebdomadal Council at Oxford University, which meets weekly.

COUNTED OUT

The intimate link between number and language is nowhere better and more vividly shown than in the many traditional children's counting rhymes that exist. Some are for 'positive' counting, to determine a choice or person or object. Others are for 'negative' counting, to decide which player shall leave a game or be eliminated. Either way, the rhymes themselves are clearly quite ancient in some instances, and appear to preserve snatches of old or half-forgotten languages.

One of the best known in the English-speaking world is (in one of its versions) as follows:

Eeny, meeny, miny, mo,
Catch a nigger by his toe.
If he hollers let him go,
Eeny, meeny, miny, mo.

The first and last line of this seem to echo French *un, deux, trois*, with *deux* acquiring the final *-n* of *un* to form the 'meeny'. (If the rhyme is really French in origin, there may be some substance in the suggestion that the second line is actually a corruption of *Cache ton poing derrière ton dos*, 'Hide your fist behind your back', as proposed by Iona and Peter Opie in *The Oxford Dictionary of Nursery Rhymes*. The rhyme could therefore be French-Canadian in origin.)

A similar rhyme has been recorded in Germany, beginning:

Eene deene Bohnenblatt,
Unser' Küh sind alle satt.

('One, two, bean and leaf, All our cows will be good beef.')

A more complex counting rhyme has been recorded in South Africa, where it is (or was) used by English-speaking children:

Une, dune, des,
Catlo wuna wahna wes,
Each, peach, muskydom
Tillatah, twenty-one.

Analysing this, the German scholar and numerologist Karl Menninger interprets the first two lines as French 'Une, deux, trois, Quatre-cinq, six-sept, huit-neuf, dix', and the last two lines as English 'Each speech must be dumb Till I come to twenty-one' (*Number Words and Number Symbols*, p. 123).

Another apparent nonsense rhyme recorded in the Midlands of England runs as follows:

Yan tan tethera pethera pimp,
Sethera lethera hovera covera dik,
Yan-a-dik tan-a-dik, bumpit,
Yan-a-bumpit, figitt.

This has been used for counting both sheep in the fields and stitches in knitting, and by adults as well as children.

What is it? The successive words can almost certainly be assigned number values, as follows:

1 2 3 4 5
Yan tan tethera pethera pimp,
6 7 8 9 10
Sethera lethera hovera covera dik,
11 12 15
Yan-a-dik tan-a-dik, bumpit,
16 20
Yan-a-bumpit, figitt.

The words need only to be compared with the Welsh numbers 1 to 20 (pp. 17 and 19) to show that this is clearly Celtic, and that it could even represent the original Celtic inhabitants of Britain who were driven westwards to Wales by the invading Anglo-Saxons in the 6th century AD.

A similar counting rhyme begins:

Eetern feetern penny pump,
All the ladies in a lump.

And there is even a suggestion of Celtic numbers in the opening words of the familiar nursery rhyme 'Hickory, dickory, dock', which could well represent 8, 9 and 10.

The seemingly meaningless jingles beloved of children may therefore not be so meaningless after all, but numerically significant—and of ancient origin.

COUNTUP ... OR COUNTDOWN?

Apart from the legacies of foreign languages quoted above, there are many familiar counting schemes and rhymes in modern English. We all remember the count to 100 with eyes closed in the game of hide-and-seek ('. . . 96, 97, 98, 99, 100!'), and have heard and seen countdowns of rockets and space vehicles ('5, 4, 3, 2, 1, Lift-off!'). In other fields, there is the bandmaster's or conductor's '3, 4' at the start of a musical piece, and the more original chant, '2, 4, 6, 8, Who do we appreciate?', followed by the name of the admired one.

But there are also longer and more substantial counts, frequently set to verse or poetry, and often with a standard accompanying tune. The count score may not be particularly high, but there is usually a memorable or enjoyable story to go with the numbers.

Basically, such 'count rhymes' are either countups or countdowns, with the enumeration progressing steadily upwards from 1 or gradually descending to it.

Here are some of the best-known counting rhymes and songs, with '[rep.]' used to denote repeated words or phrases or lines for sake of brevity. The sources of the rhymes are also given where known, together with notes on their content.

COUNTUPS

1 *The first day of Christmas*
My true love sent to me
A partridge in a pear tree.

The second day of Christmas
[rep.] *Two turtle doves, and* [etc.]

The third day
[rep.] *Three French hens,* [etc.]

The fourth day
[rep.] *Four colly birds,* [etc.]

The fifth day
[rep.] *Five gold rings,* [etc.]

The sixth day
[rep.] *Six geese a-laying,* [etc.]

The seventh day
[rep.] *Seven swans a-swimming,* [etc.]

The eighth day
[rep.] *Eight maids a-milking,* [etc.]

The ninth day
[rep.] *Nine drummers drumming,* [etc.]

The tenth day
[rep.] *Ten pipers piping,* [etc.]

The eleventh day
[rep.] *Eleven ladies dancing,* [etc.]

The twelfth day
[rep.] *Twelve lords a-leaping,* [etc.]

The Twelve Days of Christmas, traditional (meaning uncertain).

2 *I'll sing you one-O!*
Green grow the rushes-O!
What is your one-O?
One is one and all alone,

And ever more shall be so!
I'll sing you two-O!
[rep.] *Two, two, the lily-white boys*
Clothed all in green-O, [etc.]

(and so on, with the last complete verse as follows)

I'll sing you twelve-O!
Green grow the rushes-O!
What is your twelve-O?
Twelve for the twelve apostles,
Eleven for the eleven who went to heaven,
And ten for the ten commandments,
Nine for the nine bright shiners,
Eight for the April rainers,
Seven for the seven stars in the sky,
And six for the six proud walkers,
Five for the symbols at your door,
And four for the Gospel makers,
Three, three the rivals
Two, two, the lily-white boys [etc., as in verse 1]

Green Grow the Rushes-O!, also known as *The Dilly Song*, traditional. An apparent blend of religion and astrology/astronomy, but with some numbers of uncertain reference.

3 *One man and his dog*
Went to mow a meadow [rep.]

Two men went to mow,
Went to mow a meadow,
Two men, one man and his dog,
Went to mow a meadow.

(Then so on, cumulatively, up to about 'Ten men', with 'Spot' sometimes added after 'dog'.)
One Man Went to Mow, 19th century. A simple and effective count rhyme used to accompany a tour or to pass the time.

4 *One I love, two I love,*
Three I love, I say,
Four I love with all my heart,
Five I cast away;
Six he loves, seven she loves, eight both love.
Nine he comes, ten he tarries,
Eleven he courts, twelve he marries.

One I Love, 18th century. A 'divination rhyme' used by young girls when picking petals.

5 *One, two, buckle my shoe;*
Three, four, knock at the door;
Five, six, pick up sticks;
Seven, eight, lay them straight;
Nine, ten, a big fat hen;
Eleven, twelve, dig and delve;
Thirteen, fourteen, maids a-courting;
Fifteen, sixteen, maids in the kitchen;
Seventeen, eighteen, maids in waiting;
Nineteen, twenty, my plate's empty.

One, Two, Buckle My Shoe, 19th century. A popular nursery rhyme, with an apparent correlation between the numbers and a child or young person's age, and a description of increasingly complex or important tasks or roles. But the climax seems to suggest that the young person is unemployed!

6 *One for sorrow, two for joy,*
Three for a girl, four for a boy,
Five for silver, six for gold,
Seven for a secret, never to be told.

One for Sorrow, Two for Joy, 18th century. The reference is to the number of magpies seen. The wording varies considerably, especially for the higher numbers, and the rhyme can serve as a 'love verse', especially when the second line runs 'Three for a wedding, four for a birth'. In one version, 'Four' is for death, making both 1 and 4 unlucky numbers.

7 *1 and 1 are 2, that's for me and you.*
2 and 2 are 4, that's a couple more.
3 and 3 are 6, barley-sugar sticks.

LATIN NUMBERS

English basically has two kinds of number-words, cardinal (one, two, three, etc.) and ordinal (first, second, third, etc.), with a rudimentary run of numeral adverbs (once, twice, thrice). Latin had at least five types of numbers, however. They include **cardinals**, answering the question *Quot?* (How many?), **ordinals**, answering the question *Quotus?* (Which in order of number?), **distributives**, answering the question *Quoteni?* (How many each?), **numeral adverbs**, answering the question *Quotiens?* (How many times?) and **multiplicatives**, answering the question 'How many fold?'. The ordinals began *primus*, *secundus*, *tertius*, the distributives *singuli*, *bini*, *terni*, the numeral adverbs *semel*, *bis*, *ter*, and the multiplicatives *simplex*, *duplex*, *triplex*. Most of these survive as elements of English words today. The only word to incorporate *semel*, however, is the grammatical term *semelfactive*, literally 'doing once', used of a verb describing a single action performed once. Russian has some semelfactive verbs, such as *morgnut'*, 'to blink once', and *kusnut'*, 'to take a single bite'.

4 and 4 are 8, tumblers at the gate.
5 and 5 are 10, bluff seafaring men.
6 and 6 are 12, garden lads who delve.
7 and 7 are 14, young men bent on sporting.
8 and 8 are 16, pills the doctor's mixing.
9 and 9 are 18, passengers kept waiting.
10 and 10 are 20, roses—pleasant plenty!
11 and 11 are 22, sums for brother George to do.
12 and 12 are 24, pretty pictures, and no more.

Christina Rossetti, *1 and 1 are 2*, 19th century. A rhyme for children, as a pleasant variation on the all too familiar 'Twice 2 are 4' or cumulative multiplication tables.

COUNTDOWNS

1 *There were ten green bottles,*
Standing on the wall,
Ten green bottles,
Standing on the wall,
And if one green bottle
Should accidentally fall,
There'd be nine green bottles,
Standing on the wall.

(And so on, with the number gradually decreasing.) *Ten Green Bottles*, traditional, 19th century. Commonly used to accompany a tour or walk.

2 *Ten little nigger-boys went out to dine;*
One choked his little self, then there were nine.
Nine little nigger-boys stopped up very late;
One overslept himself, then there were eight.
Eight little nigger-boys travelling in Devon;
One said he'd stay there, then there were seven.
Seven little nigger-boys chopping up sticks;
One chopped himself in half, then there were six.
Six little nigger-boys playing with a hive;
A bumble-bee stung one, then there were five.
Five little nigger-boys going in for law;
One got in chancery, then there were four.
Four little nigger-boys going out to sea;
A red herring swallowed one, then there were three.
Three little nigger-boys walking in the zoo;
A big bear hugged one, then there were two.
Two little nigger-boys sitting in the sun;
One got frizzled up, then there was one.
One little nigger-boy left all alone;
He got married, and then there were none.

Ten Little Niggers, song written by Frank Green, 1869, for the Christy Minstrels. The song was almost certainly based on the American *Ten Little Injuns* (see below). As 'nigger' is now a term of ethnic abuse, the song is seldom if ever heard, and Agatha Christie's novel of 1939, *Ten Little Niggers*, was renamed *Ten Little Indians* when her dramatized version of it opened in New York in 1944.

3 *Ten little Injuns, standing in a line,*
One toddled home and then there were nine.
Nine little Injuns, swinging on a gate,
One tumbled off and then there were eight.
Eight little Injuns, gayest under heaven,
One went to sleep and then there were seven.
Seven little Injuns, cutting up their tricks,
One broke his neck and then there were six.
Six little Injuns, kicking all alive,
One kicked the bucket and then there were five.
Five little Injuns, on a cellar door,
One tumbled in and then there were four.
Four little Injuns, up on a spree,
One he got fuddled and then there were three.
Three little Injuns, out in a canoe,
One tumbled overboard and then there were two.
Two little Injuns, fooling with a gun,
One shot t'other and then there was one.
One little Injun, living all alone,
He got married and then there were none.

NUMBER CRUNCHER

Paul Erdös, who celebrated his 75th birthday in 1988, has been described as the world's most prolific mathematician. He travels ceaselessly, having no home, no family, no money and no pursuit other than his obsessive search for elegant solutions to mathematical problems. Born a Hungarian Jew, Erdös is said to have thought about more problems than any other mathematician in history. He has written or co-authored over 1000 papers, and in 1987 alone published 50 papers, more than most mathematicians write in a lifetime. For Erdös, mathematics is order and beauty at its purest, and he remains obsessively fascinated by prime numbers. At the age of 17, as a college freshman, he caused a stir in mathematical circles by coming up with a simple proof that there is always a prime number between any integer greater than 1 and its double. When asked once why numbers held such a spell over him, he replied, 'If you don't see why, someone can't tell you. I know numbers are beautiful. If they aren't beautiful, nothing is.'

Ten Little Injuns, comic song by S. Winner, 1868. The song was immediately popular, producing many variants, such as *Ten Little Niggers* (see above). Other versions were: *Ten Little Negroes*, *Ten Little Darkies*, *Ten Youthful Africans*, and a parody, *Ten Undergraduates*, which began: 'Ten undergraduates going out to dine, One got proctorised, then there were nine.' These were all current in the 1860s.

4 *There were ten in the bed and the little one said*
Roll over, roll over;
They all rolled over and one fell out,
There were nine in the bed and the little one said
Roll over, roll over; [etc.]

Ten in the Bed, popular song, 20th century (?). A song still sung at scout camps, 'singsongs' and the like. Like the three other countdowns above, the gradual diminution proceeds from 10 down to 0.

3 FROM SECOND TO CENTURY

THE ORIGINS OF UNITS OF TIME AND THE CALENDAR

The basic unit of time measurement was originally, as it still is, the **day**. It was usually reckoned from dawn to dawn, or sunrise to sunrise, and the Babylonians and Greeks measured their day thus, while the Egyptians reckoned their day from midnight to midnight and the Jews from sunset to sunset. Later, the Germanic races counted the passage of time by nights, rather than days, hence the modern 'fortnight' ('fourteen nights'), not 'fortday'.

The day itself was divided in different ways. The Sumerians subdivided it into six watches, three in the daytime, three at night, and the Jews observed a similar division. But most of western civilization came to adopt a 24-hour division, with 12 hours for the day and 12 for the night. This particular figure (2 × 12 = 24) was based on the Sumerian sexagesimal (60-based) system of reckoning, which was itself based on gradations of 60 (5 × 12 = 60), although the 24-hour day was devised by the Egyptians.

The Christian church adopted its own division of the day for daily worship, with seven 'day hours': Lauds, Prime, Terce, Sext, None, Vespers and Compline. These so called 'canonical hours' evolved in the 4th century, and the arrangement itself was determined by St Benedict.

For most people, however, the day was divided into standard **hours** as we still have them, with each hour subdivided into 60 **minutes**, and each minute further subdivided into 60 **seconds**.

The very names 'minute' and 'second' derive from their origin as a division and subdivision of time. The minute was the *pars minuta prima*, or 'first minute part' of the hour (and also of the degree), while the second was the *pars minuta secunda*, or 'second minute part' of the hour.

LARGER TIME UNITS

Once the day was divided, the next task was to group the days in suitable reckoning units. The obvious grouping was the number of days in a **month** or lunation, with this in turn subdivided into four **weeks** representing (approximately) the four phases of the Moon: new, first quarter, full and last quarter. Each of these phases lasts roughly seven days, which accorded with the existing Babylonian interest in the symbolic number 7. Moreover, the Jews had adopted the seven-day week by the 1st century BC, and this in turn had its influence on Christendom and the Christian calendar. Not only was the world created in seven days (six 'active' + one rest day), but the seven days themselves linked up with the seven planets, which were then regarded as the Sun, the Moon, Mars, Mercury, Jupiter, Venus and Saturn.

Hence, from this last, the names of the seven days of the week in many languages: Sunday

PROPOSED WORLD CALENDAR

In 1888 the French astronomer Gaston Armelin proposed a universal calendar of 12 months divided into four quarters with 91 days in each, each month containing 26 workdays. An extra day would be added at the end of the year every year, and a second extra day would be added after 30 June every leap year. The calendar has yet to be officially adopted worldwide, however, if only because it would disrupt existing religious festivals.

	January April July October	February May August November	March June September December
S	1 8 15 22 29	- 5 12 19 26	- 3 10 17 24
M	2 9 16 23 30	- 6 13 20 27	- 4 11 18 25
T	3 10 17 24 31	- 7 14 21 28	- 5 12 19 26
W	4 11 18 25 -	1 8 15 22 29	- 6 13 20 27
T	5 12 19 26 -	2 9 16 23 30	- 7 14 21 28
F	6 13 20 27 -	3 10 17 24 -	1 8 15 22 29
S	7 14 21 28 -	4 11 18 25 -	2 9 16 23 30*

* Intercalated day after 30 December every year; plus second intercalated day after 30 June in a leap year

after the Sun, Monday after the Moon, Tuesday after Mars (as in French *mardi*), Wednesday after Mercury (as French *mercredi*), Thursday after Jupiter (French *jeudi*), Friday after Venus (*vendredi*) and Saturday after Saturn. In English, the names of Norse gods were substituted for four of these (Tiw, Tuesday; Woden, Wednesday; Thor, Thursday; Frigg, Friday).

THE YEAR AND ITS MONTHS

The question next arose of fixing an even greater time unit. As the lunation of the Moon (the period of approximately 29½ days in which it orbited the Earth) was used to reckon the month, so the time taken by the Earth to move round the Sun (or the Sun round the Earth, as was then supposed) could serve to give a longer period, one of about 365 days. This **year**, as it was, could then be subdivided into 12 months, with this particular figure again based on the sexagesimal count (12 is a divisor of 60).

The lunation of the Moon provided a suitable monthly time unit in other ways, too. Not only was it short enough to provide a smallish and conveniently calculated number of days (about 29), but it was also quite close to the natural menstrual period of a woman. ('Menstrual' derives from Latin *mensis*, 'month'.)

The problem was: how many days exactly should each month have? The Babylonians alternated months of 29 and 30 days, while the Egyptians kept to 30, as did the Greeks, who adopted their reckoning.

THE ROMAN CALENDAR

The Romans, however, evolved a more complex system, basing it on a calendar system that was observed in what is now Italy in about 1000 BC. This earlier calendar was an agricultural one, based on the natural period of growth of plants and crops and the annual times of year traditionally reserved for sowing and harvesting.

On the Apennine peninsula (modern Italy), where the Roman calendar evolved, the climate is subtropical, giving an agricultural (or vegetational) year of about 300 days. Roman astronomers therefore used this figure as a basis when working out their own calendar, at the same time taking into account the fact that the lengths of day and night vary throughout the year but that there are two days in the year when the day equals the night. These two days fall on the vernal (spring) and autumnal equinoxes, respectively 21 March and 23 September.

The Romans thus compiled a calendar with an agricultural year of 304 days, dividing this into 10 months. For the beginning of the year, they naturally chose that month in which the vernal equinox occurred (21 March). There are still some countries today, such as Iran, where the year begins at this time.

Each of the 10 Roman months was given a name to indicate its order and position in the year, as follows:

I	Primidilis	VI	Sextilis
II	Duodilis	VII	September
III	Tridilis	VIII	October
IV	Quartidilis	IX	November
V	Quintilis	X	December

The first four months had 31 days and the remainder 30. But the year ended at the onset of winter, leaving an apparently uncounted gap (two months, by our reckoning) before commencing again in the spring. To fill this gap, the Emperor Numa Pompilius (who is semi-legendary) is said to have added the two months of January and February, and to have added 50 days to the existing calendar, making a total of 354. He then apparently deducted one day from the 30-day months, leaving 51 to be shared between January and February. But as the Romans had a superstitious dread of even numbers, January was given an extra day. February was thus still left with an even number of days. But this was regarded as appropriately inauspicious for the infernal gods, to whom it was dedicated. This system allowed the new 12-month year to have 355 days, an odd number.

However, the calendar year was out by several days on the solar year, so in an attempt to put things right a further 'month' of 22 or 23 days called Mercedonius (or Intercalaris) was inserted or 'intercalated' between 23 and 24 February. This intercalation was so clumsily done, that by the time of Julius Caesar the civic or calendar year was even further out, and was now about three months ahead of the solar.

In his capacity as Pontifex Maximus, therefore, Caesar intercalated enough days to bring the year 46 BC to a total of 445 days. This was thus to be the *ultimus annus confusionis*, 'last year of muddled reckoning'.

From the next year on, the Egyptian solar calendar was adopted for Roman use by inserting enough days in the shorter months to bring the total up to 365 and to arrange for an extra day, not a 'month', between 23 and 24 February every leap year. This extra day was called *bissextus*, 'bissextile'.

For the purposes of the standardized 12-month year, the first four months had new names to replace their former 'number names'. Primidilis became Martus (after Mars, the god of war, and also of agriculture), Duodilis became Aprilis, perhaps from *aperire*, 'to open', referring to the budding plants, Tridilis became Maius, after Maia, goddess of mountains and fruitfulness, and Quartidilis became Junius, after Juno, goddess of abundance.

Later, Quintilis was renamed Julius after Julius Caesar, who was born in this month, and Sextilis was renamed Augustus, after his adopted son Octavianus, whose title was 'Augustus' ('the great').

ROMAN DAYS AND MONTHS

Dates in the Roman calendar were expressed according to three key days of the month. These were the Calends (*Kalendae*), the 1st of the month, the Nones (*Nonae*), the 9th, and the Ides (*Idus*), the 13th. However:

In March, July, October, May
The Nones fall on the 7th day
And the Ides on the 15th

Julius Caesar thus had to beware 15 March, the 'Ides of March'.

All other dates fell a given number of days before (*ante*) one of the key days, so that 23 January, for example, was the 10th day before the Calends of February, otherwise *ante diem decimum Kalendas Februarias*, abbreviated to *a.m. X Kal. Feb.*

'Bissextile' is the modern word to describe a leap year. It derives from Latin *bis sextus*, 'twice sixth', as this was the name given to the extra day added in a leap year before 24 February in the Roman calendar. 24 February itself was *ante diem sextum Kalendas Martias* (*a.m. VI Kal. Mart.*) so that the intercalated day repeated this 6th day, which thus came twice.

FROM JULIUS TO GREGORY

Caesar's calendar, known as the Julian calendar, remained in European use until the 16th century. By then, however, it had once again become too long. It was based on a length of 365.25 days, whereas the true length of the solar year (the precise time taken for the Earth to revolve round the Sun) is 365.242199 days. This error of 11 minutes 14 seconds a year had resulted in the vernal equinox falling on a day 10 days different from its true date by 1545, and it was regarded as important to correct this error, since the vernal equinox (21 March) was used to determine Easter.

No correction was made, however, until 1582, when Pope Gregory XIII issued a papal bull to put the calendar right. The Feast of St Francis, 5 October, was redesignated as 15 October, thus omitting ten days. Moreover, to bring the year closer to the true solar year, a value of 365.2422 days was accepted. But as even this would result in an error of 3.12 days every 400 years, it was decreed that three out of four centennial years should not be leap years unless they were divisible by 400, the standard definition of a leap year being that it is divisible by 4. This meant that the three centennial years 1700, 1800 and 1900 were not leap years, and that the 3-day error was made good. The year 2000 *is* divisible by 400, however, so it will be a leap year.

This new Gregorian calendar is the one that

most western countries follow today, although by no means all converted from Julian to Gregorian at the same time, and England, for example, adopted the New Style calendar (as it is also known) only in 1752.

NUMBER NAMES IN OTHER CALENDARS

Apart from the Roman calendar, other calendars have subsequently evolved with number names for the months of the year or the days of the week.

One of the more original was the French Republican calendar, introduced in 1793, which not only renamed the months and days, but rearranged the existing Gregorian calendar to form a new year comprising 12 months of 30 days each, with each month in turn divided into three *décades* or 10-day 'weeks'. This rearrangement left five intercalary days (six in a leap year) which were called *sans-culottes*, 'breechless ones', after the French Revolutionaries' own name for themselves, as wearers of long trousers instead of the hitherto fashionable knee-breeches. The five extra days were observed as national holidays.

The chronology of the Republican calendar was the work of a French mathematician, Gilbert Romme. The months were renamed after natural phenomena, such as Germinal, 'seed-time', and Fructidor, 'fruit-time', while the 10 days in each *décade* were given ordinal number names: primidi, duodi, tridi, quartidi, quintidi, sextidi, septidi, octodi, nonidi, décadi, this last corresponding to Sunday and being a day of rest.

The Republican calendar, which was antedated to begin on 22 September 1792, the day of the Proclamation of the Republic, lasted for 12 years, 2 months, 27 days, and on 1 January 1806 the Gregorian calendar was reintroduced.

There are still calendars with number names today in other countries of the world. One of the best-known with monthly number names is the Japanese, with 12 months as follows: Ichigatsu, Nigatsu, Sangatsu, Shigatsu, Gogatsu, Rokugatsu, Shichigatsu, Hachigatsu, Kugatsu, Jugatsu, Ju-ichigatsu, Ju-nigatsu. In these, *ichi*, *ni*, *san*, etc. are the numbers 1, 2, 3, etc. while *gatsu* is 'month'.

In general, however, it is the days of the week rather than the months that have number names, with the day count running either from Sunday or Monday, and not necessarily with all seven days having a number name.

Differing from other Romance languages, Portuguese has a Sunday-to-Saturday count, as follows: domingo, segunda-feira, terça-feira, quarta-feira, quinta-feira, sexta-feira, sabado. In the names for Monday to Friday, feira equals 'day' (literally 'fair', i.e. trading day).

In Arabic, the days are also counted from Sunday, with *yaum* meaning 'day': yaum al-ahad ('the one'), yaum al-itnain ('the two'), yaum at-talâta ('the three'), yaum al-arba'a ('the four'), yaum al-khamîs ('the five'), yaum al-juma'a ('the uniting'), yaum as-sabt ('the sabbath').

In modern Greek, the count likewise begins on Sunday: Kyriaké ('Lord's day'), Deutéra ('Second'), Tríte ('Third'), Tetárte ('Fourth'), Pémpte ('Fifth'), Paraskeué ('Preparation'), Sábbato ('Sabbath').

In Chinese, however, the count begins on Monday, with the days from Sunday running: xīngqīrì ('sun'), xīngqīyī ('one'), xīngqī'èr ('two'), xīngqīsān ('three'), xīngqīsì ('four'), xīngqīwu ('five'), xīngqīliù ('six'). In all these, *xīnqī* represents 'day' (but actually means 'week', literally 'star time').

Most Slavonic languages have retained number names only for Tuesday, Thursday and

NEW STYLE CONVERSIONS

Different countries converted from the Julian calendar to the Gregorian, or New Style, at different times, as follows:

1582 Italy, Spain, Portugal, France, Poland, Holland
1584 Austria, Switzerland
1587 Hungary
1610 Prussia
1700 Denmark, Norway
1752 England, Ireland
1753 Sweden
1873 Japan
1875 Egypt
1912-17 Albania, Bulgaria, China, Romania, Turkey, Yugoslavia
1918 Soviet Russia
1923 Greece
1967 Vietnam

Countries converting between 1582 and 1700 adjusted by 10 days, between 1700 and 1800 by 11 days, between 1800 and 1900 by 12 days, and after 1900 by 13 days (which is the figure that will apply until the year 2100).

Friday. In Russian, for example, the days from Sunday run as follows: voskresen'ye ('Resurrection'), ponedel'nik ('day after Sunday'), vtornik ('second day'), sreda ('middle day'), chetverg ('fourth day'), pyatnitsa ('fifth day'), subbota ('Sabbath').

A WORLD CALENDAR?

In order to avoid the various irregularities of the Gregorian calendar, with its months of unequal lengths (28, 29, 30 or 31 days) and a second half of the year that can vary by two or three days from the first half, attempts have been made to evolve a simpler calendar which could be universally adopted.

In 1923 a special committee was set up under the auspices of the League of Nations to consider the matter. After the Second World War the project was then assigned to the Economic and Social Council of the United Nations.

Many proposals have been put forward, with a basic choice between calendars of 12 and 13 months. A 13-month calendar was proposed in 1849 by the French philosopher Auguste Comte (1798–1857). According to his plan, every month would have 28 days and be divided into four weeks, with each month beginning on a Sunday and ending on a Saturday. However, 13 × 28 = 364, leaving an extra day in the year. Comte proposed that this day would be nameless, but would be intercalated after the final Saturday of the 13th month before the New Year began, when it would be an extra holiday. In a leap year, an extra day would also be added after the last Saturday of the 6th month.

But a 13-month year would have its problems, if only because when the year is divided into quarters, the months would have to be divided, too.

Attention has therefore been concentrated on a new 12-month calendar proposed in 1888 by the French astronomer Gaston Armelin. This calendar would comprise 12 months divided into four quarters with 91 days in each. The first month of the year would have 31 days, and the rest 30. The first day of the first month would fall on a Sunday, and each quarter would end on a Saturday and have 13 weeks. A month would contain 26 working days. In a non-leap year, an extra day would be added at the end of the year, after 30 December, as a Year-End Day or World Day, while in a leap year a second extra day would be intercalated after 30 June.

This World Calendar has been approved by several countries, including France, India, Yugoslavia and the USSR, but has not as yet been finally approved by the UN. One objection has been made on religious grounds, since if the new calendar were introduced, the intercalated day or days would disrupt the steady succession of days of the week, and this would interfere with the religious observance of many faiths, where festivals are associated with a particular number of days and with particular days of the week. The 'twelve days of Christmas', for example, would become 13!

Even so, the proposed calendar has practical advantages over the existing Gregorian calendar with its many variables.

THE 21ST CENTURY

When will the 21st century begin? Not on 1 January 2000! A simple calendar calculation will show that this is so. The 1st century began on 1 January AD 1, and the 100 years making that entire century ended on 31 December AD 100. The 2nd century thus began the following day, 1 January 101. The 19th century therefore ended on 31 December 1900, and the 20th century began on 1 January 1901. The 20th century will therefore close on 31 December 2000, so that the 21st century will open on 1 January 2001.

THE MEASUREMENT OF TIME

Conventionally, and universally, time is measured in hours, minutes and seconds, by a variety of timepieces, from clocks and watches to more sophisticated devices such as the twin atomic hydrogen masers (Microwave Amplification by Stimulated Emission of Radiation) at the US Naval Research Laboratory in Washington, DC, at present the world's most accurate time measurer.

Two familiar time measurers that deviate from the norm, however, are the **ship's bell** and the **metronome**.

THE SHIP'S BELL

A seaman's day is still officially divided into watches, with the 24-hour period comprising seven watches, five of four hours each and two of two hours. The latter, shorter periods are designed to vary the time and period of the watch for a seaman working alternate watches.

The times of the watches, together with their names, are:

12 noon to 4 p.m.	Afternoon Watch
4 p.m. to 6 p.m.	First Dog Watch
6 p.m. to 8 p.m.	Second Dog Watch
8 p.m. to 12 midnight	First Watch
12 midnight to 4 a.m.	Middle Watch
4 a.m. to 8 a.m.	Morning Watch
8 a.m. to 12 noon	Forenoon Watch

To indicate the passage of time through these watches, the ship's bell is rung at half-hourly intervals. For the five four-hour watches, one bell is rung at the end of the first half-hour, two bells at the end of the first hour, and so on up to eight bells at the end of the watch.

For the two dog watches of two hours each, four bells mark the end of the first dog watch, and eight bells the end of the second (or last) dog watch. One traditional exception to this, however, is that one single bell is rung after the end of the first half-hour of the second dog watch (i.e. at 6.30 p.m.) instead of five. This has been the practice since the mutinies of 1797 in the Royal Navy, because five bells at this time was to be the signal for mutiny. Officers discovered this plan in one port, and ordered one bell to be rung instead, thus preventing the mutiny.

THE METRONOME

The metronome (whose name means 'measure rule') marks the passage of quite a different kind of time, that of the tempo of a piece of music.

The instrument itself produces a clear clicking sound from a pendulum whose swing can be adjusted to indicate the desired speed of performance for a particular piece. Most metronomes are still clockwork instruments, and some have a bell to strike at every second, third or fourth beat as well as the clicking pendulum to mark the beats themselves.

The best-known metronome is the one devised by the German inventor Johann Nepomuk Maelzel (1772–1838): hence the initials 'M.M.' that are still found at the start of a piece of music, together with an indication of the actual tempo, for example, 'M.M. ♩ = 100', meaning that the metronome must be set to beat 100 times a minute, with each beat representing a crotchet (or quarter-note).

Many famous composers added their own metronomic tempi to their compositions, to give an indication of the speed at which they should be played. Some of Schumann's markings are extremely fast, however, suggesting that his own metronome was functioning incorrectly.

THE 24-HOUR CLOCK

The 24-hour clock is in fairly common and especially formal use to mark a particular time of the day or night. Its aim is to avoid ambiguity, as '7.00' can mean 7.00 a.m. or 7.00 p.m.

The 24-hour period is marked thus, compared to the conventional 'a.m.' and 'p.m.' times:

12 midnight	0000 (or 2400)
1 a.m.	0100
2 a.m.	0200
3 a.m.	0300
4 a.m.	0400
5 a.m.	0500
6 a.m.	0600
7 a.m.	0700
8 a.m.	0800
9 a.m.	0900
10 a.m.	1000
11 a.m.	1100
12 noon	1200
1 p.m.	1300
2 p.m.	1400
3 p.m.	1500
4 p.m.	1600
5 p.m.	1700

MONTHLY FIRSTS

1 January New Year's Day
1 February Abolition of Slavery Day
1 March St David's Day
1 April April Fool's Day
1 May May Day
1 June Glorious First of June
1 July Canada Day
1 August Lammas
1 September Partridge Day (partridge may be shot)
1 October Pheasant Day (pheasants may be shot)
1 November All Saints' Day
1 December Three Jesuit Martyrs

6 p.m.	1800
7 p.m.	1900
8 p.m.	2000
9 p.m.	2100
10 p.m.	2200
11 p.m.	2300

The 24-hour clock is mainly found in use today in the Armed Services, in public transport timetables, in TV and radio programme schedules (although not in Britain), and in the majority of digital clocks and watches.

The hour on the 24-hour clock can be expressed verbally in different ways, however, so that 1600, for example, can be simply 'sixteen hundred', or 'sixteen hundred hours' (which can also be written 1600 hrs), or 'sixteen double-O'. In this last way of speaking, the time 0400 would be 'O-four-double-O'.

NAMES LIKE NUMBERS

If the first day of the week is regarded as Monday, it will be found that the English names of the days are curiously similar to their corresponding cardinal numbers, especially if Wednesday is regarded as the 'midweek' day, as in other Germanic languages (such as German *Mittwoch*):

Monday/One Day
Tuesday/Two Day
Wednesday/Midst Day
Thursday/Four Day
Friday/Five Day
Saturday/Six Day
Sunday/Seven Day

'Saturday' also happens to be similar to 'Sabbath Day', which is what it is.

WRITING THE DATE

UK and US conventions vary slightly when writing the date.

Britons usually write the date as '22 December 1989', while Americans tend to prefer 'December 22, 1989'.

When the date is written entirely in numbers, care may be needed to determine the month, depending on the nationality of the writer.

To a Briton, 9/10/89 is 9 October 1989, but to an American it is 10 September 1989, because the month comes first.

Some people, especially European continental writers, like to express the month in Roman figures, so that 9 October 1989 would be 9/x/89, or 9-x-89. This at least avoids any ambiguity of month.

Historic dates near the year 0 are expressed by AD (*anno Domini*, 'year of the Lord') after 0 and by BC (before Christ) before 0, thus: AD 352, 65 BC.

4 STYLES AND SINGULARITIES

GROUPS AND TYPES OF NUMBERS

The study of numbers involves the use of several technical terms. Some, such as *cardinal number* and *ordinal number*, are usually familiar to the average reader (although even these two are sometimes confused). Others, such as *deficient number* and *rational number*, will probably need some explanation. Others again, such as *amicable number* and *weird number*, to say nothing of *emirps* and *repunits*, seem decidedly strange by their mere designations alone. Here, therefore, is a brief guide to some of the 'professional' types of number that are found.

abundant number an even number that is less than the sum of its **factors** (excluding itself). Example: 12, whose factors (1 + 2 + 3 + 4 + 6) add up to 16. (Compare **deficient number**.)

algebraic number a number that appears as the solution of an algebraic equation. Example: 5, appearing as the result of the equation $3x + 1 = 16$ (where it is represented by x).

amicable number one of a pair of numbers having **factors** that add up to form the other number of the pair. The lowest amicable numbers are 220 (an **abundant number**) and 284 (a **deficient number**). Factors of 220 are 1 + 2 + 4 + 5 + 10 + 11 + 20 + 22 + 44 + 55 + 110: total 284. Factors of 284 are 1 + 2 + 4 + 71 + 142: total 220.

automorphic number a number any power of which (such as its **square** or **cube**) ends in itself. Examples: 5 (squared = 25) and 76 (cubed = 438,976). 'Automorphic' literally means 'self form'.

cardinal number the most familiar type of number, used in counting (1, 2, 3, 4, 5, etc.), and so named as it is a 'hinge' number (Latin *cardo, cardinis*, 'hinge') on which all other types 'swing' or depend. (Compare with **ordinal number**.)

complex number a 'two-dimensional' (planar) number that consists of a 'real part' and a so-called 'imaginary part'. It always involves the **square root** of -1 (expressed by the letter i), and typically appears in the form $2 + 3i$, where 2 is 'real' but $3i$ is regarded as 'imaginary'. In fact it is not imaginary at all, but planar rather than linear, like most numbers, since multiplication by i implies rotating the number point through 90° anticlockwise, i.e. from so many units 'east' to so many units 'north'.

composite number any number that is not a **prime number**, such as 4 or 6 or 49.

cube a number raised to the third power, that is, multiplied by itself twice. Example: 343, which is 7^3, which is $(7 \times 7 = 49) \times 7$.

cyclic number a number which, when multiplied by any number from 1 to the number that is the total of its digits, always consists of the same digits as the original number, appearing in the same cyclic order but beginning at a different point. The usual example quoted is 142,857:

$1 \times 142{,}857 = 142{,}857$
$2 \times 142{,}857 = 285{,}714$
$3 \times 142{,}857 = 428{,}571$
$4 \times 142{,}857 = 571{,}428$
$5 \times 142{,}857 = 714{,}285$
$6 \times 142{,}857 = 857{,}142$

deficient number any number that is not an **abundant number**, that is, one whose **factors** add up to a total lower than itself, such as 14, whose only factors are $1 + 2 + 7 = 10$. As most numbers have very few factors, there are far more deficient numbers than abundant.

emirp a **prime number** that turns into another prime number when it is reversed ('emirp' is 'prime' backwards). Examples: 13 and 31, 17 and 71. (See also **palindromic number**.)

factor 1 a number which, when multiplied with another, forms a third number (the product), such as 5 and 6 in the equation $5 \times 6 = 30$; 2 any number that divides into another number, such as 5, which divides 6 times into 30. In this book, 'factor' is used in the second sense. The word literally means 'doer'. (Compare **factorial**.)

factorial a number that is the product (total) of all the numbers from 1 to some other given number multiplied together. Example: 24, which is the product of $1 \times 2 \times 3 \times 4$, and so is called 'factorial 4'. The symbol for 'factorial' is '!', so that 'factorial 4' is written '4!', colloquially called either '4 bang' or '4 shriek'. Factorials increase very rapidly, so that 20!, for example, is already a 19-figure number (2,432,902,008,176,640,000). 'Factorial' simply means 'involving **factors**'.

Fibonacci number a number in the sequence 0, 1, 1, 2, 3, 5, 8, 13, 21, 34, 55, 89, 144, 233, and so on, in which every number after the second is the sum of the two preceding numbers. The sequence, which is infinite, is named after the 15th-century Italian mathematician Leonardo Fibonacci, who discovered it. The sequence is widely found in nature, so that the arrangement of leaves spirally up a stem, for example, is usually such that the number of leaves between two positions where a leaf lies precisely over another on the stem is a Fibonacci number.

irrational number a number that cannot be expressed in terms of one whole number being divided by another, as most numbers can. Examples: any **square root** (such as $\sqrt{2}$) and pi (symbol π), as the ratio of the circumference of a circle to its diameter. When expressed in decimals, irrational numbers go on for ever without repeating, so that pi is (or begins) 3.141592653598 . . . . In fact, irrational numbers are so unconventional that they have to be represented by special symbols, not by familiar numerals. (Compare **rational number**.)

logarithm the number, written as a superscript (small number to the rop right of another), expressing the power to which a fixed number (the 'base') must be raised to obtain a third number. In the expression $b^x = N$, for example, if b is the base and equal to 10, and N is a number, equal to 100, then x is equal to 2 and is said to be the logarithm of 100 to the base 10. This is written: $\log 100 = 2$ (in which it is understood that 'log' means 'logarithm to the base 10'). (Put another way, this expresses the fact that 10^2, or 10 squared, = 100.) Logarithms are used to simplify multiplication and division, and are (or were, before the advent of the pocket calculator) employed by being looked up in tables. The word itself means literally 'reckoning number'.

natural number a name for any of the **cardinal numbers**, or 'counting' numbers (1, 2, 3, 4, 5, 6, 7, etc.), with which we are most familiar. Natural numbers are also known as integers.

normal number a lengthy number, such as an

THE NUMERICAL SCIENCES

The names of some of the most familiar numerical sciences can sometimes have unexpected origins.

Mathematics, the science of magnitude and number and related topics, derives from Greek *mathema*, 'learning'.

Arithmetic, the science of numbers and reckoning, derives from Greek *arithmetike tekhne*, 'art of counting'.

Algebra, the science of calculating by symbols, takes its name from Arabic *al-jabr*, 'the reunion', a term originally applied to the resetting of broken bones, but later used of anything broken, hence of the combination or 'reintegration' of numbers.

Geometry, the area of mathematics that deals with the properties of points, lines, surfaces and solids, has its origin in Greek *geometria*, literally 'earth measurement'.

Trigonometry, the branch of mathematics that deals with the sides and angles of triangles, takes its name from Greek *trigonon*, 'triangle' and *metron*, 'measure'.

Calculus, the system of calculation used in higher mathematics, gets its name from Latin *calculus*, 'pebble', as early calculations were visually worked with pebbles. The verb **calculate** is directly related to this, so has the same origin.

irrational number, in which every numeral (1, 2, 3, 4, etc.) occurs as frequently as any other. For example, in the first one million digits of pi, there are about 100,000 ones, 100,000 twos, and so on.

ordinal number any of the numbers used to indicate order (hence the name), such as 1st, 2nd, 3rd, 4th, 5th, etc. (Compare **cardinal number**.)

palindromic number a number which reads the same backwards as forwards, whether the numerals are identical (such as 111,111) or vary mirror-fashion either side of a central numeral (such as 12,321 or 2,867,682). In a sense all numbers from 1 to 9 are palindromic. Such numbers are not simply satisfyingly symmetrical to look at, but also have another special property. If you take any number and then add to it the same number with its digits reversed, continuing the process as necessary, you sooner or later finish up with a palindromic number. Sometimes the palindrome comes quickly, as in 18 + 81 = 99. Other palindromes take more steps, such as 68 + 86 = 154; 154 + 451 = 605; 605 + 506 = 1111. The number less than 100 which takes the most steps is 89, which needs 24 'sums' before it finally produces the palindrome 8,813,200,023,188. The term itself means literally 'running back', and is familiar also in language (to describe words such as 'bib' and 'reviver').

SATISFYING SYMMETRIES

Some quite simple sums produce pleasing patterns if they are worked successively. The following three are quoted by Joseph Madachy in his *Mathematics on Vacation* (1966):

A

$1 \times 8 + 1 = 9$
$12 \times 8 + 2 = 98$
$123 \times 8 + 3 = 987$
$1234 \times 8 + 4 = 9876$

B

$3 \times 37 = 111$
$6 \times 37 = 222$
$9 \times 37 = 333$
$12 \times 37 = 444$

C

$1^2 = 1$
$11^2 = 121$
$111^2 = 12321$
$1111^2 = 1234321$

The reader is invited to continue each of these progressions and see what happens.

perfect number a number that is the sum of all its possible **factors**, excluding itself. Example: 28, which is the sum of 1 + 2 + 4 + 7 + 14. The first four perfect numbers are 6, 28, 496 and 8128, and all perfect numbers end either in 28 or in 6 preceded by an odd numeral. An example of the latter is the fifth perfect number: 33,550,336. To date, only 30 perfect numbers have been calculated, the 30th having 130,099 digits. It was recorded on a CRAY supercomputer in 1985 in Houston, Texas, when the largest known **prime number** was also calculated.

polygonal number one of a sequence of numbers formed when a polygon (many-sided figure), real or imaginary, is formed with 2, 3, 4, 5, etc. points along its sides and its 'middle' filled accordingly. The simplest polygonal sequence is formed from triangular numbers, using an equilateral triangle, like this:

```
*    *      *       *        *
    * *    * *     * *      * *
          * * *   * * *    * * *
                 * * * *  * * * *
                         * * * * *
```

This figure generates the sequence 1, 3, 6, 10, 15, 21, 28, 36, 45, 55, 66, ... A pentagon formed like this would generate the sequence 1, 5, 12, 22, 35, 51, 70, 92, . . . , and hexagonal numbers would be 1, 6, 15, 28, 45, 66, 91, ... Put another way, triangular numbers are formed by adding every numeral from 1 successively to produce the sequence:

$$1 = 1$$
$$1 + 2 = 3$$
$$1 + 2 + 3 = 6$$
$$1 + 2 + 3 + 4 = 10$$
$$1 + 2 + 3 + 4 + 5 = 15$$
$$1 + 2 + 3 + 4 + 5 + 6 = 21$$
$$1 + 2 + 3 + 4 + 5 + 6 + 7 = 28$$

Pentagonal and hexagonal numbers would successively add the third and fourth numbers after 1 respectively, i.e. 1, 1 + 4, 1 + 4 + 7, etc. and 1, 1 + 5, 1 + 5 + 9, etc. (See **square number** for the missing numeral here.)

prime number a number that can be divided exactly only by itself and 1. The lowest prime numbers are 2, 3, 5, 7, 11, 13, 17, 19, 23, 29, 31 and 37. Their frequency gradually, but irregularly, decreases the higher one goes. Apart from the first prime number (2), all primes are odd numbers, not even. The highest prime number yet calculated (on a CRAY supercomputer in Houston, Texas, in 1985) has 65,050 digits, and is mathematically expressed as $2^{216091}-1$. The computer took three hours to check that this number was indeed prime, working at the rate of 400 million calculations per second. The record result was broadcast by the BBC at 7.30 a.m. on 18 September 1985. Any new prime number calculated like this automatically results in a new **perfect number**.

random number not just any number, chosen at random, but (mathematically, anyway) a number that cannot be described more compactly than by simply stating its digits in order. Example: 298,859,200,066,702,864,083,897. (This number was actually arrived at by a mathematician taking balls numbered 1, 2, 3, 4, 5, 6, 7, 8, 9, 0 and picking them out of a hat, returning each one before making the next choice.) Put negatively, random numbers have nothing special about them, and there is no way in which they can be concisely described or in which they can be calculated. The number of digits in a random number is, of course, infinite.

rational number a number produced by dividing one whole number by another, even when the result is a fraction. All whole numbers are themselves therefore rational numbers. Examples: 3 (6 divided by 3) and ½ (3 divided by 6). (Compare **irrational number**.)

reciprocal number any number that results from 1 being divided by the stated number, and so that is related to it reciprocally (as a mathematically derived 'alternative'). Examples: the reciprocal of 3 is 1 divided by 3 which is 0.3333333 . . . (that is, 0.3 recurring, or indefinitely), and the reciprocal of 2 is 0.5. One special feature of reciprocals is that the sum of the reciprocals of all the **factors** of a **perfect number** (including, for once, the number itself) is always exactly equal to 2. Example: the factors of 6 are 1, 2, 3 and 6; $(1/1) + (1/2) + (1/3) + (1/6) = 2$.

repunit the abbreviated term (pronounced as if two words, 'rep unit', in full 'repeated unit') for a number that consists entirely of units, or the digit 1, but not including 1 itself. The smallest repunit is thus 11, and the second is 111. Repunits can be manipulated to form interesting numbers or pretty patterns. Example: $1{,}111{,}111{,}111{,}111^2$ (i.e. squared) = 12,345,678,900,987,654,321. The term 'repunit' was invented by the American mathematician Albert Beiler.

sociable number one of a 'chain' of numbers whose **factors** add up to the next number in the chain, with the process repeated until a number is reached whose own factors add up to the first number in the chain. Example: the factors of 12,496 add up to 14,288, whose own factors add up to 15,472, whose own factors add up to 14,536, whose own factors add up to 14,264, whose own factors add up to 12,496 (which was where we started). This is therefore a 5-link chain of sociable numbers. It was the first to be discovered, in 1918, and it was the only such chain (apart from a 28-link one) known until 1969, when computers were brought in to search for sociable groups with 10 links or fewer for all numbers up to 60 million. Seven new chains were found, each having four links, but despite diligent searching, no chain of sociable numbers has yet been found with just three links. But they may exist, and they have been provisionally named 'crowds'.

square number a number in a series of squares (whole numbers multiplied by themselves) that can be visually and mentally associated with the **polygonal numbers**, since the series can be represented pictorially by increasingly large squares, each containing the number of units in the number. The squares of 1, 2, 3, 4 and

RECORDS, RECORDS

What is the highest number or the lowest?

Obviously, there is no actual limit to which a man or machine can count. All that is lacking is a suitable name to express the extreme figure reached.

The highest lexicographically accepted number in the system of successive powers of ten is the *centillion*. But outside the decimal system the highest yet recorded and named is the Buddhist *asankhyeya*, equal to 10^{140}, and mentioned in Jainist writings of the 1st century BC.

The lowest number has not yet been definitively agreed. Logically, it is 0, which is a number so low that it is non-existent. But the smallest fraction or the largest minus number would also be the smallest number, in its own terms. Clearly, either of these, like positive (plus value) numbers, could be extended equally to infinity. As the English mathematician Augustus de Morgan wrote: 'Great fleas have little fleas upon their backs to bite 'em, And little fleas have lesser fleas, and so *ad infinitum*.'

5 are thus 1, 4, 9, 16 and 25, and they can be shown thus:

```
*    * *    * * *    * * * *    * * * * *
     * *    * * *    * * * *    * * * * *
            * * *    * * * *    * * * * *
                     * * * *    * * * * *
                                * * * * *
```

Ancient Greek mathematicians were fascinated by numbers which could be displayed by arranging points in regular patterns like this, either on a plane or in space. The converse of a square is a square root, otherwise that number which, when multiplied by itself, makes a given number. The square root of 9 is thus 3, while the square root of 8 is 2.8284271 . . . . (A square that is a whole number, like 4 and 9, is a 'perfect square'.) The conventional symbol for square root is $\sqrt{}$, so that '$\sqrt{2}$' means 'square root of 2' (which is 1.4142135623730950488016887242096980785697 . . .).

transcendental number any number that is not an **algebraic number**, that is, one that cannot appear as the solution of an algebraic equation. The first transcendental number was found by the French mathematician Joseph Liouville in the mid-19th century. It was a decimal fraction full of zeros beginning 0.110001000000000000000001000 . . . . Pi (π) is now also known to be a transcendental number. And in a sense almost *all* numbers are transcendental. This is because if a piece of string is cut arbitrarily into two, it is extremely unlikely that the two pieces would have lengths that could be measured in algebraic numbers! Transcendental numbers are so called because they 'transcend', or go beyond, algebraic numbers.

untouchable number a number that is never the sum of the **factors** of any other number. The sequence of untouchable numbers starts: 2, 5, 52, 88, 96, 120 . . . .

weird number an **abundant number** that does not represent the total of any addition of its own **factors**. The smallest weird number is 70. The factors of this are 1, 2, 5, 7, 10, 14 and 35, which when added together come to 74. This makes it an abundant number. But no combination of the factors will sum to 70. It is therefore a weird number. Weird numbers are rare, and there are only seven of them below 10,000. They are: 70, 836, 4030, 5830, 7192, 7912 and 9272. It should be noted that they are all even numbers, and this may be a condition of weird numbers. At any rate, no one has yet been able to find a weird number that is an odd number.

BEYOND NUMBER

The Avatanshaka Sutra, the chief scripture of Mahayana Buddhism in India, contained one teaching that had great appeal. This was that fundamentally there is no number, 'neither one nor two'. The idea was that once we enter in thought into the region beyond duality, that is, into the 'Buddha-mind', multiplicity disappears. But so does unity or 'oneness'. In short, we have reached beyond any type of classification, including that of numbers.

NUMERICAL PECULIARITIES

Many numbers have interesting associations or properties, and in a few cases unusual historical backgrounds. Some of the most important features connected with individual numbers from 1 to 1000 are given below.

1 The unity number, serving as a 'base' for all the others. The Greeks regarded 1 (wrongly) as both odd and even, because when added to an odd number, it produces an even, and when added to an even, it produces an odd. It has no factors, and does not alter any other number that it multiplies. It is not normally regarded as a prime number, even though it is divisible by no other number but itself and 1.

2 Long regarded as a number representing many human characteristics and natural phenomena, and seen by the Greeks as a number that has a 'beginning' and 'end' but no 'middle'. $2 + 2 = 2 \times 2$, which gives it a unique arithmetical status. It is the first prime number, and the only *even* prime. It is also the first deficient number. Importantly, it serves as the basis for the binary system.

3 One of the most important numbers in religion and mysticism and directly associated with the triangle, which appears in many guises as a mystic symbol. Division into 3 is common not only in the natural world, but in language, which has three degrees of comparison (positive, comparative, superlative), three genders (masculine, feminine, neuter), three persons (first, second, third) and (in some languages) three numbers (singular, dual, plural). A number is divisible by 3 if its digits add up to a multiple of 3. It is the second prime number, and the first odd prime. In the field of factorials: 3 = 1! + 2!.

4 The first composite number and the second square number (2^2). It is 'big' in the natural world, with four cardinal points of the compass and the 'four corners of the earth', as well as the four winds. The Greeks believed in four elements: earth, air, fire, and water. The number has long been associated with solid objects, especially a cube, which has square faces.

5 The smallest automorphic number, with all powers of 5 ending in the digit 5: when multiplied by itself, it ends in itself (5 × 5 = 25). Although a potentially suitable number to serve as a base for a counting system (five fingers), it is known to be used for this purpose only in one language, Saraveca, spoken in South America in the Arawakan family of languages.

6 The second composite number and the first with two distinct factors. Euclid defined 6 as the first perfect number, as it is the sum of its factors 1 + 2 + 3. (It is moreover the only number that is the sum of *exactly* three of its factors.) Its square ends in 6 ($6^2 = 36$), and the only other number to have a square ending in itself is 5.

7 A number, like 3, that has many religious and mystic associations. It is a powerful 'calendar' number, associated with the number of days in a week and the number of days in a lunar month (4 × 7).

8 The second cube number ($2^3 = 8$), the first being 1. A number closely associated with music, with 8 notes in an octave. If the numbers 1 to 10 are written in alphabetical sequence, 8 will come first in several languages, including English, German and Russian.

9 The third square number ($3^2 = 9$), and the only square that is the sum of two consecutive cubes ($1^3 + 2^3 = 9$). In factorials, 1! + 2! + 3! = 9. A number can be divided by 9 only if its digits add up to 9. The first nine numbers make a unique magic square.

10 The familiar base of our counting and reckoning system and basis of the metric system of weights and measures, with quantities either multiplied or divided by 10. It is also the sum of the first four numbers (1 + 2 + 3 + 4 = 10), and can be represented visually as a triangle to show these numbers:

The Greeks called this figure the *tetraktys*, or 'number four'.

11 The smallest repunit, and so, like all repunits, divisible by the product of its digits (1 × 1 = 1). It is a palindromic prime number, and is (or was) an important multiple in the imperial system of length measures (e.g. 22 yards = 1 chain, 220 yards = 1 furlong, 1760 yards = 1 mile).

12 The first abundant number, and one strongly associated with the calendar and time generally (12 hours in a day, 12 months in a year, 12 signs of the zodiac). It is the long familiar basis of the duodecimal system of reckoning, beloved of Britons, for whom there were 12 pence to a shilling and 240 pence to a pound, and for whom there still are 12 inches to a foot. It is the only number to have an alternative name as itself ('dozen') and as its square ('gross'). It is divisible by the sum of its digits (1

π

Probably the best-known and most unusual of all numbers is π (pronounced 'pie'), the ratio of the circumference of a circle to its diameter. It is the only irrational and transcendental number that occurs naturally. Here it is to 40 places of decimals: 3.1415926535897932384626433832795028841972...

Ever since the Babylonians became interested in π, from about 2000 BC, there have been increasingly accurate but lengthy determinations of its value. Here are some of them, with their degree of accuracy:

Mathematician	*No. of decimal places*
Archimedes (3rd century BC)	(2—used fractions)
Ptolemy (2nd century AD)	(4—used fractions)
Al-Kashi (15th century)	16
Ludolph von Ceulen (16th century)	35
John Machin (1706)	100
William Shanks (1853)	707 (with, however, an error after place 528)
ENIAC computer (1949)	2037
CDC 6600 computer (1967)	500,000
Tamura and Kanada (1983)	16,777,216

The Greek letter π itself stands for *peripheria*, meaning 'circumference'.

$+ 2 = 3$) and by their product ($1 \times 2 = 2$). Not only does its square (12^2) = 144, but this total reversed, produces an accurate answer: $441 = 21^2$. In the 20th century, 12 has become associated with the 12-tone musical scale, known also as the dodecaphonic scale. This itself is otherwise the conventional 'chromatic scale', ascending from C by semitones. (The 12 notes are C, C♯, D D♯, E, F, F♯, G, G♯, A, A♯, B.)

13 The (in)famous unlucky number, one greater than the near 'perfect' 12. Even so, there are 52 (13×4) weeks in a year, and 13 cards in each suit of a standard pack: Ace, 2, 3, 4, 5, 6, 7, 8, 9, 10, Jack, Queen, King, with the ace able to rank as highest as well as lowest.

14 Associated both with weights and measures (14 pounds = 1 stone) and the calendar (14 days = 1 fortnight).

15 Associated with the smallest magic square, whose rows, columns and diagonals add up to 15. Also associated with sports, such as rugby (15 players), snooker (15 balls) and tennis (15 as the first scoring point).

16 The fourth square number ($4^2 = 16$) and the only number that is the perimeter ($4 + 4 + 4 + 4$) and the area (4×4) of the same square. 16 is also the basis for the hexadecimal system used by computers, denoted by the numerals 0 to 9 and the six letters A to F. As such, hexadecimal can be easily converted to binary, simply because $16 = 2^4$.

17 One of only six numbers equal to the sum of the digits of its cube ($17^3 = 4913$, and $4 + 9 + 1 + 3 = 17$). (The other five such numbers are 1, 8, 18, 26 and 27, of which three are themselves cubes.)

18 Familiar to schoolchildren for the reversible sums: $9 + 9 = 18$, and $9 \times 9 = 81$. 18 is also equal to the sum of the digits of its cube: $18^3 = 5832$, and $5 + 8 + 3 + 2 = 18$.

19 A number with few straightforward properties, and a 'near miss' to a round or productive number, as it is 1 short of 20 and as $1 + 9 = 10$.

20 A significant number in systems of counting and of weights and measures, with 20 shillings formerly in a pound and 20 itself serving as the basis for the vigesimal reckoning, based on the score (twoscore = 40, threescore = 60, fourscore = 80, etc.). In weight, 20 hundredweight = 1 ton.

24 A familiar calendar number, with 24 hours in the day. In weights and measures, too, 24 scruples = 1 ounce and 24 grains = 1 pennyweight. It is the smallest number to have factors whose product is a cube: $1 \times 2 \times 3 \times 4 \times 6 \times 8 \times 12 = 13{,}824 = 24^3$. It is also a famous factorial: 4!.

25 A number that is both a square (5^2) and the sum of two squares ($3^2 + 4^2 = 25$).

27 A number that is the sum of the digits of its own cube: $27^3 = 19{,}683$, and $1 + 9 + 6 + 8 + 3 = 27$.

28 The number of days in a lunar month, and the number of pounds in a quarter in the imperial system of weights and measures. 28 is also the second perfect number, after 6.

30 A calendar number, with four of the 12 months having 30 days. An hour, too, is frequently divided as two half-hours of 30 minutes each.

31 Like 30, a calendar number, with seven of the 12 months having 31 days. 31 is also the third perfect number as the sum of its divisors $1 + 2 + 4 + 8 + 16$. These are also doubling or squared powers, and the only other known number that can be written in two ways as the sum of successive powers like this is 8191, which is both $1 + 2 + 4 + 8 + 16 + 32 + 64 + 128 + 256 + 512 + 1024 + 2048 + 4096$ and $1 + 2 + 2^2 + 2^3 \ldots + 2^{12}$.

39 A number which appears to have no special properties and so to be quite uninteresting—which makes it interesting!

40 A number that constantly recurs in the Bible as a 'full' number.

47 A number interesting for, if nothing else, producing two sums, the second with a reversed total and a changed sign: $47 + 2 = 49$, and $47 \times 2 = 94$.

50 Half a hundred, of course, and recognized as such by the Roman letter L. 50 is also the product of the two smallest numbers that are themselves the sum of two squares: 50 = 5 (which is $1^2 + 2^2$) $\times$ 10 (which is $1^2 + 3^2$).

52 The number of weeks in a year, and also the number of cards in a standard pack (without jokers).

60 The basis of sexagesimal counting. Hence the division of a circle into 360 degrees, and the division of a degree into 60 minutes. On the clock, and more familiarly, we thus equally have an hour of 60 minutes, and a minute of 60 seconds. These are the only common international measurements that have not been metricated.

69 As a curiosity, 69 is the only number whose square and cube together use all the digits from 0 to 9 once each: $69^2 = 4761$, and $69^3 = 328{,}509$.

70 The smallest weird number.

81 Not only 3^4, but the only number whose square root equals the sum of its digits (apart from the insignificant 0 and 1): $\sqrt{81} = 9 = 8 + 1$.

88 Not only a repeated digit, but a number whose square has two repeated digits: $88^2 = 7744$.

90 Familiar as the number of degrees in a right angle.

99 A number that is 1 off 100 and that is itself linked with the number 1 in various ways. For example, as a decimal fraction, $1/99 = 0.010101010101\ldots$, and $9 \times 11 = 99$.

100 One of the most familiar 'big' round numbers, and the square of 10, which is the base of the decimal system. An old brainteaser is to link the digits 1 to 9 by various mathematical signs in order to produce a total of 100. One of the most straightforward answers is: $1 + 2 + 3 + 4 + 5 + 6 + 7\ (8 \times 9) = 100$. One of the shortest answers, using only three signs, is: $123 - 45 - 67 + 89 = 100$.

120 Not only 2×60 but, factorially, 5!.

121 A palindromic square of a palindrome ($11^2 = 121$), and also the only square that is the sum of consecutive powers from 1: $1 + 3 + 9 + 27 + 81 = 121$.

128 A power of 2 whose own digits are also powers of 2. It is not known whether 128 is the only number to have all its digits as powers of 2 in this way. In the binary code, 128 is 10,000,000.

132 The smallest number to equal the sum of all the two-digit numbers made from its own digits: $12 + 21 + 13 + 31 + 23 + 32 = 132$.

136 Add together the cubes of this number's digits ($1^3 + 3^3 + 6^3$) and you get 244. Add together the cubes of *this* number's digits ($2^3 + 4^3 + 4^3$) and you have the number you started with, 136.

144 A gross, or a dozen dozens, or 12^2, and the equivalent of '100' in the duodecimal system of counting.

180 The number of degrees in a half-circle (i.e. twice two right-angles of 90° each) and also the sum of the angles of any triangle. 180 also happens to be the number of degrees Fahrenheit between the freezing point of water (32°F) and its boiling point (212°F).

216 The smallest cube (6^3) that is also the sum of three cubes: $3^3 + 4^3 + 5^3 = 216$.

256 A number that is 1,000,000 in binary (because it is 2^8) and 100 in a hexadecimal (16-based) system. Both systems are used in computers.

360 The number of degrees in a full circle, and approximately the number of days in a year, if divided into 12 months of 30 days each. The zodiac was originally divided into 360, with each of the 12 signs subdivided into 30 equal parts.

484 The palindromic square of a palindromic square root: $22^2 = 484$.

500 A half-thousand, and recognized as such by the Romans, who assigned 500 the letter D.

666 A favourite number with numerologists and lovers of the occult, as it is the 'Number of the Beast'. In Roman numerals it uses the six Roman letters once only, DCLXVI, which may have influenced its special status.

729 The second smallest cube (9^3) to be the sum of three cubes: $1^3 + 6^3 + 8^3 = 729$. It is also the sum of five cubes: $1^3 + 3^3 + 4^3 + 5^3 + 8^3 = 729$. This is because $3^3 + 4^3 + 5^3 = 6^3$ (= 216). This in turn means that in base 3, 729 is 1,000,000. The Greeks regarded 729 as a special number, as it is the square of 27, which is the cube of 3. (If the powers of 2 and 3 are given to 9, $1 + 2 + 3 + 4 + 8 + 9$, their sum is 27.)

873 A factorially interesting number, as $1! + 2! + 3! + 4! + 5! + 6! = 873$.

999 All the digits from 1 to 9 can be combined in three three-figure numbers to make this total: $149 + 263 + 587 = 999$.

1000 A familiar 'very big' round number that will be 10^3 in any base at all. Although 'K' (from Greek *khilioi*, 'thousand') is commonly used as an abbreviation for 1000, especially in job salaries (e.g. £20K), 1K of memory in a computer is 1024. This is because computers count in binary, and 2 to the power of 10 (2^{10}) = 1024.

5 MAGIC AND MYSTIC NUMBERS

NUMEROLOGY, ANCIENT AND MODERN

Like astrology, numerology is a pseudo-science. Like it, too, it is an ancient art, and one that still has many practitioners and adherents today. Moreover, the two sciences are mutually related, and from the earliest times people have linked numbers with the stars, combining calculation and observation. The heavenly bodies, after all, are not merely numbered but have a frequency of motion (or apparent motion), and our current calendar, as explained, has evolved precisely from a blending of the two disciplines of 'star-study' and 'number-study'.

Without going into intricate detail, it can be said that numerology largely concerns itself with people's birth dates, as much the day as the year, and that it has formulated simple rules for calculating a particular person's 'lucky' number, or the number that would be suitable for any partner. There is nothing new in this, for the ancient Greeks and Babylonians also related number and person in a similar fashion, adding the digits of a compound number to obtain a single number (1 to 9).

TOLSTOY'S SPECIAL NUMBER

The great Russian novelist, Leo Tolstoy, regarded 28 as his special number. He was born in 1828 on 28 August (old style). His eldest son, Sergei, was born on 28 June. Just before he died, Tolstoy left his home at Yasnaya Polyana for the last time on 28 October. When he was asked to select poems by French modernist poets for his treatise *'What is Art?'*, Tolstoy deliberately picked the poems from page 28 of different poetry books, so as not to be accused of bias. In his novel *Resurrection*, the hero, Prince Nekhlyudov, experiences his spiritual revival on 28 April. And some unpublished extracts from his short story *The Kreuzer Sonata* were found to contain these words: 'I was then young and in love. I was 28, and was going to visit my future wife's family to propose to her.'

The calculation of a person's lucky or special number is a simple affair. If you were born on any day from the 1st to the 9th of the month, that day is your lucky number. For higher dates, you merely add the two digits. A person born on the 13th, for example, has a lucky number of 4 (1 + 3). In some instances, such an addition will have to be made twice. A person born on the 29th, for example, adds 2 and 9 to obtain 11, then 1 and 1 to obtain 2, which is the lucky number, or 'spirit number', as some numerologists like to call it.

In a similar fashion, it is possible to derive a special number from the letters of a person's name, using a different number-value for each letter. The basis of figuring can vary. The simplest (recommended by the numerologist Juno Jordan) is the following, each letter having the value of the number above it:

1	2	3	4	5	6	7	8	9
A	B	C	D	E	F	G	H	I
J	K	L	M	N	O	P	Q	R
S	T	U	V	W	X	Y	Z	

Another, followed by the famous numerologist Cheiro, and believed by him to be of ancient origin, is as follows:

1	2	3	4	5	8	5	5	1	2	3	4	5	7	8
A	B	C	D	E	F	G	H	I/J	K	L	M	N	O	P
1	2	3	4	6	6	6	5	1	7					
Q	R	S	T	U	V	W	X	Y	Z					

The fact that there is no 9 in the figuring is because this number was said to represent the 9-lettered name of God, so could have no single letter assigned to it. But this need not prevent 9 or a 9-based number from appearing in the calculation.

Just as the digits of a compound number in a person's birth date are continuously added until they produce a single number, so this process

MYSTIC 153

In the Bible, *John* 21:11 tells how the disciple Simon Peter netted 153 fishes. This number was inevitably interpreted mystically by the early church fathers, and in particular by St Augustine, who explained the figure as follows:

Apud vos numerate: sic computate. Decem et septem faciunt centum quinquaginta tres: si vero computes ab uno usque ad decem et septem et addas numeros omnes—unum, duo, tria: sicut unum et duo et tres faciunt sex; sex, quattuor et quinque faciunt quindecim: sic pervenes ad decem et septem, portans in digitis centum quinquaginta tres.

Count for yourselves: reckon as follows. 10 and 7 make 153: if you count from 1 to 17 and add all the numbers—1, 2, 3: just as 1 and 2 and 3 are 6; 6, 4 and 5 are 15: so you arrive at 17, carrying on your fingers 153.

10, of course, was the number of the Ten Commandments in the Old Testament, while in the New Testament 7 is frequently found as a mystic number.

St Augustine simply discovered that if you add the numbers 1 to 17 (itself 10 + 7), you get 153. What better way of expressing the number's special significance?

can be used for a person's name, thus (using Juno Jordan's table):

J A N E
1 1 5 5 = 12 1 + 2 = **3**

A M A N D A
1 4 1 5 4 1 = 16 1 + 6 = **7**

B R O W N
2 9 6 5 5 = 27 2 + 7 = **9**

3 + 7 + 9 = **19** 1 + 9 = **10** 1 + 0 = **1**.

So that Jane Amanda Brown is fortunate enough to have the highly desirable *name number* of 1. If she chooses, she can also obtain a *birth number* by adding the digits of not just her birth date, but her day, month and year of birth, thus:

March 22 1973
3 + **4** + (1 + 9 + 7 + 3 = 20; 2 + 0 = 2) **2** = **9**

She can then finally add both propitious name number (1) and birth number (9) to get her *reality number* (or whatever she likes to call it), so that 1 + 9 = 10, and 1 + 0 = 1, which brings her back to her highly advantageous 1 again. Magic!

Numerologists think of the correspondence of letters and numbers like this as a **'vibration'**, since they purport to designate a person's life, both spiritual (or 'inner') and physical. Letter and number are regarded as thus being rhythmically in tune with each other, of 'living', in fact.

But as well as being associated with a particular person, a number is also thought of as having an affinity with a particular **colour** and **day of the week** (and also **jewel** or **stone**), making that colour, day and stone the favoured one for the numbered individual.

Some of the mainstream numerological associations of certain numbers with particular human characteristics are given below, together with their related weekday and colour (but not stone), which in turn are associated with a number's own planet (in the astrological rather than the astronomical sense).

Sceptical readers are likely to remain sceptical. But no book on numbers can overlook the field of numerology.

MAGIC NUMBERS

1 1 has always been a special, literally unique number. The Greeks believed that 1 denoted

13—STILL UNLUCKY

'Bramshott and Liphook Parish Council has decided not to increase its membership to 13, because members felt the number was unlucky. Councillors heard they were entitled to one extra councillor because of the increased population of the parish—but opted to wait until they could have 14 members.' (News item in local paper, *Petersfield Post*, 4 January 1989.)

the spirit from which the whole visible world was created. 1 combined male and female, and was reason, good, harmony and happiness on the one hand, and matter, darkness and chaos on the other.

The uniqueness of 1 exists mathematically, too. Five of its remarkable features are:

(a) It is neither a prime number nor a composite number;
(b) Every natural number can be divided by it;
(c) It is the only natural number which has one divisor—itself;
(d) Any number multiplied or divided by 1 is exactly the same after the operation as before;
(e) Any number apart from zero, if divided by itself, will always produce the answer 1.

In numerology, 1 represents the **Sun**, and stands for everything that is creative, inventive and positive. 1 people are ambitious and seek to 'reach the top' in all they do. Their most favoured day is Sunday, and their favoured colour is gold or yellow, or some shade of it, such as bronze or golden brown.

2 2 is the source of all inequality and contradiction. It represents opinion, which itself embodies falsehood and truth. As an even number, 2 is regarded as female, while odd numbers are male. Yet 2 evolved from 1, and is the immediate 'child' of it.

In numerology, 2 represents the **Moon**. 2 people are gentle, imaginative and romantic, with qualities more mental than physical. They harmonize well with 1 people, however, and to a lesser extent with 7 people. Their favourable days are Monday, or Friday (governed by Venus), and their favourable colour is green or white.

3 Just as 2 and 1 go together, so do 2 and 3, and as the only pair of consecutive simple numbers they are unique. 3 is also the only natural number that is the sum of all preceding numbers. Its symbol is the triangle. In religion and folklore, 3 is frequently found. In ancient Babylonia 3 main gods were worshipped: Sun, Moon and Venus. In ancient Egypt there were also 3 gods: Horus, Osiris and Isis. For the Greeks, 3 was a divine number and a symbol of perfection in all natural objects and phenomena. The Romans worshipped 3 main gods also: Jupiter, Mars and Quirinus. For Christians, 3 symbolizes the Holy Trinity: God in 3 persons, Father, Son and Holy Spirit. In classical literature, there were 3 Fates, 3 Graces and 3 Furies, as well as 9 (3 times 3) Muses. Today, we repeat oaths 3 times ('the truth, the whole truth, and nothing but the truth') and we give 3 cheers. In the New Testament, Peter denied Christ 3 times.

In numerology, 3 represents the planet **Jupiter**, and it is strongly linked to any multiple of itself, but especially to 9. Like 1 people, 3 people are ambitious. They aim to rise to the top and gain authority. In achieving this, however, many 3s become dictatorial, and make enemies. 3 people, too, are proud, and markedly independent. They fret when frustrated or thwarted. Their favourable day is Thursday, and their favourable colour is mauve or purple, with blue and crimson also good, but as secondary shades.

4 The Greeks regarded 4 as a symbol of all that is known and unknown, a quality it acquired from being 3 (the living world) plus 1. But the 4 mathematical functions of add, subtract, divide and multiply also made 4 a symbol of natural objects, which evolve and develop by adding, subtracting, multiplying or dividing. The Greeks, too, held that 4 was a symbol of strength. This is because the sum of the first 4 numbers (1 + 2 + 3 + 4) equals 10, or perfection. Moreover they noticed that with these first 4 numbers it is possible to arrive at all other single figures up to 10: 1 + 4 = 5, 2 + 4 = 6, 3 + 4 = 7, 1 + 3 + 4 = 8, 2 + 3 + 4 = 9. Visually, 4 was the symbol of the pyramid, with its 3 sides and base. In the Old Testament, there were 4 rivers of Paradise, and in the New Testament there are 4 gospels. In ancient and medieval times, it was believed that a person's temperament was determined by 4 'humours': phlegm, blood, choler and black bile.

In numerology, 4 represents the planet **Uranus**. 4 people are consistently different: they oppose standard opinions, rebel against authority, and are generally unconventional. They do not make friends easily, and are on the whole less successful in life than many other numbers. One of their favourable days is Saturday, and their favourable colours are half-shades or half-tones, such as grey or beige.

5 5, like 7, has long been a significant 'step' number, not simply because of the 5 fingers of the hand, but as the sum of the lowest even number (2) and the lowest odd number (3). (1 is not an odd number since it does not have 1 left over when divided by 2, the definition of 'odd'.) As the sum of 2 and 3, 5 is therefore a symbol of marriage, for 2 is female and 3, as an odd number, is male.

In numerology, 5 represents the planet **Mercury**, and 5 people are thus 'mercurial' and highly strung. This means that they can be either quick in thought and impulsive in action, or generally resilient in character, with an ability to recover rapidly from any shock or blow. But although they can be good, they can also be bad, and if the latter, they are likely to remain so. On the plus side, they make friends easily, and are keen speculators and risk-takers, which can turn them into successful money-makers. Their fortunate days are Wednesday and Friday, and their favourable colours are any shade of grey or white. However, their gregarious nature enables them to harmonize with most colours satisfactorily, although light shades are better than dark.

6 6 stands for equilibrium, as it represents two triangles, base to base. It has long been regarded as a 'perfect' number, if only because God created the universe in 6 days.

In numerology, 6 represents the planet **Venus**, and 6 people have a 'magnetic' quality that attracts others to them. Bound in with this is their romantic nature: they love and are loved. Equally, they are artistic, and their popularity makes them generous entertainers and hosts.

RING OUT THE WRONG . . .

According to an ancient tradition, the Japanese ring the bells of the Buddhist temples 108 times on New Year's Eve. The number is not a random one. The Japanese believe that all people are subject to six vices: malice, stupidity, thoughtlessness, greed, indecision and covetousness. Each of these six vices has 18 shades of application. Hence the number of strokes on the bell (18 X 6), with each stroke banishing each shade of each vice. After the 108th stroke, the Japanese go to bed knowing that they can start the New Year with a real 'clean sheet'

URSULA 1:VIRGINS 1

In the Christian church, 21 October is still sometimes observed as the feast of St Ursula and the Eleven Thousand Virgins. According to medieval legends, Ursula was a British princess who went to Rome accompanied by 11,000 maidens, and who was killed with them by the Huns on returning from the pilgrimage. But such a high figure seems suspect, and none of the names of the virgin martyrs has been recorded. It is possible, however, that Ursula had 11 companions, and that an abbreviated Latin text referred to them as 'XI MV'. This was intended to represent *undecim martyres virgines*, 'eleven virgin-martyrs'. But the 'M' seems to have been understood as the Roman numeral 1000 (*mille*), so that the abbreviation was wrongly read as *undecim milia virgines*, 'eleven thousand virgins'.

However, according to another account, Ursula had only *one* companion, whose name was Onesima, from the Greek word for 'useful' (an appropriate name for a handmaiden). The affectionate form of this name is Onesimilla, which could have become distorted over the years to Latin *undecim milia*.

They are loyal to the death, and do their utmost to promote happiness and concord and to banish jealousy and discord. One of their favourable days is Tuesday, and their favourable colour is any shade of blue.

7 7 has the overall 'charisma' that 6 somewhat lacks, and 'mystic 7' abounds in many ancient human manifests, such as religious rites or archaeological figures. Some have connected 7 with the cult of the Moon, with a month divided into the two 7s of a fortnight or the four 7s of a week. At the same time, 7 is the sum of 3 and 4, two influential numbers in their own right. The force of 7 is particularly strong in the Bible, and nowhere more so than in the Book of Revelation, where mention is made of, among others, 7 candlesticks, 7 stars, the 7 churches of Asia, 7 trumpets, 7 spirits before God's throne, 7 horns, 7 vials, 7 plagues, a 7-headed monster, and the Lamb with 7 eyes. For many more famous 7s, see p. 111.

In numerology, 7 represents the planet **Neptune**, and its association with the Moon ties it closely to the number 2. 7 people are independent and individualistic, loving travel and the acquisition of knowledge generally. They can tend to be philosophers on the one hand, but effective business people on the other. They have a natural interest in religion and the occult, and are often intuitive. They share Monday as their favourable day with 2s, while their favourable colour, similarly, is green or white.

8 The Greeks regarded 8 as a symbol of death. They evolved this concept from the study of 8 when multiplied with other numbers. They then added the integers of the total to produce another answer, as follows:

$1 \times 8 = 8$	8
$2 \times 8 = 16$	$1 + 6 = 7$
$3 \times 8 = 24$	$2 + 4 = 6$
$4 \times 8 = 32$	$3 + 2 = 5$
$5 \times 8 = 40$	$4 + 0 = 4$
$6 \times 8 = 48$	$4 + 8 = 12$
$7 \times 8 = 56$	$5 + 6 = 11$
$8 \times 8 = 64$	$6 + 4 = 10$

As can be seen, the final figure gradually diminishes from 8 (produced from $1 \times 8 = 8$) down to 4 (from $5 \times 8 = 40$). But then the chain is broken, with a two-figure total appearing, and in order to continue the sequence it was necessary to add the integers of the second total to get the final figure, so that $1 + 2 = 3$, $1 + 1 = 2$, and $1 + 0 = 1$. This way, the final figure for the first 8 multiplications descends from 8 to 1. But what happens then? Continue the '8 times table' with $9 \times 8 = 72$, add these figures ($7 + 2 = 9$), and the final figure leaps up to 9, only to descend again as before. This process then continues up to $18 \times 8 = 144$ (with $1 + 4 + 4 = 9$), after which the whole cycle starts again. The final total of each calculation thus constantly diminishes or 'dies'.

In numerology, 8 represents the planet **Saturn**. 8 people are often lonely and misunderstood, with their lives either a success or a marked failure. The number is generally an unfavourable one, resulting in sorrow, loss and humiliation. But Saturday (the day of Saturn) can be a good day for 8s, while their favourable colour is black or some other dark shade.

Overall, therefore, 8 has little good to offer, despite being 4 + 4, the product of two good numbers.

9 For those peoples who have followed a lunar calendar, 9 has been a highly favourable number, as the period between two identical phases of the Moon is 27 days, and a third of this is 9. As a mystic number, 9 was believed to represent wisdom, knowledge, education and fate, and it is found in many religious contexts. Not surprisingly, the Greeks regarded 9 as a number standing for stability, valour and resilience, if only because of its unique 'self-regenerating' quality: whatever number 9 is multiplied by, the sum of the figures in the answer will always be 9. Thus, $2 \times 9 = 18$, and $1 + 8 = 9$; $9 \times 9 = 81$, and $8 + 1 = 9$; $11 \times 9 = 99$, and $9 + 9 = 18$, so that $1 + 8$, as before, $= 9$. This characteristic will always be found.

In the culture and folklore of the world, 9 is richly represented. For example, there were 9 Muses, Lars Porsena swore by 9 gods, the Hydra had 9 heads, and in Shakespeare's *Macbeth*, the 3 witches ('weird sisters') sing 'Thrice to thine, and thrice to mine, and thrice again to make up nine'. In the Bible, however, 9 is not represented as frequently as one might expect.

In numerology, 9 represents the planet **Mars**. 9 people are 'fighters', with determination and a keen desire to be their own boss. This makes them good soldiers in battle or good leaders in society, but their grim determination can also be the cause of their downfall, and they can make many enemies in their 'battle for life'. They resent criticism, too, and can have an exaggerated opinion of themselves. At the same time, they will do almost anything to win love and affection, to the extent that they will often look foolish in the process. 9s have Tuesday as their fortunate day, and any shade of red as their fortunate colour.

10 For obvious reasons (10 fingers, the first number in a new stage of reckoning), 10 has long been regarded as a symbol of harmony, completion and perfection. A genuine 'ten out

FORTY DAYS AT SIXES AND SEVENS

In the Christian calendar, *Quadragesima* is now the name of the first Sunday in Lent. Originally, it was a name for Lent itself, referring to the 40 weekdays from Ash Wednesday, when Lent begins, to Holy Saturday (Easter Even), when it ends. In this sense, *Quadragesima* (the Latin word for '40th') is accurate, for Ash Wednesday is the 40th weekday before Easter Day, counting backwards from Holy Saturday. *Quinquagesima* ('50th') is the name of the Sunday before Lent, i.e. the one previous to *Quadragesima*, and the second and third Sundays before Lent are known respectively as *Sexagesima* ('60th') and *Septuagesima* ('70th'). But these names are misnomers, and despite attempts to reconcile the stated number of days, it is more than likely that the three Sundays before Lent were named simply in numerical sequence on an analogy with *Quadragesima*.

of ten' number, therefore. The Greeks saw 10 thus, and believed that 10 heavenly bodies revolved round a 'central fire'. To achieve this magic number, they added a new heavenly body, Antiearth, to the existing fixed stars, to the Earth, and to what they regarded as the seven 'unfixed stars' or planets, Saturn, Jupiter, Mars, Venus, Mercury, Sun and Moon. All these bodies were arranged in 10 spheres, which produced a harmonious sound in motion, the 'music of the spheres'. This Antiearth later acquired the Latin name of 'Primum Mobile', or 'first moving thing'. As the outermost of the spheres, furthest from the Earth (according to the Ptolemaic concept), 'Primum Mobile' carried the other nine round with it.

11 The ancients regarded 11 as an unlucky number. This may have had something to do with the fact that the Sun, which directly affects people's health, has variations in its activity every 11 years.

12 Long regarded as an 'upper limit' number by many peoples, 12 and its duodecimal reckoning remain today as a firmly established quantitative measure. There are thus 12 months in the year, 12 hours in the day, and 12 inches in a foot, to name some of the most familiar usages. 12 was a lucky number in Babylonia, China, Rome and elsewhere, and was symbolic of sufficiency, completion and perfection. In the Bible, 12 occurs several times, such as the 12 tribes of Israel in the Old Testament and the 12 disciples of Christ in the New. In astronomy, there are 12 signs of the Zodiac, and in Arthurian legend, there were 12 Knights of the Round Table.

13 A number as unlucky as 12 is lucky, and probably so by very virtue of the fact that it follows it, and that it lacks almost all of the satisfying features (such as the many factors or divisors) that 12 has. For the ancient Hebrews, 13 was additionally unlucky because alphabetically it was designated by the letter M (a position it still holds in the English alphabet as the 13th letter), which began the word *mem*, meaning 'death'. In the Christian world, 13 is still connected with the number present at the Last Supper, and especially with Judas Iscariot, who was one of the 12 disciples at table with Jesus, and the one who would betray him for 30 pieces of silver. Even today, there are many people who consciously avoid anything connected with 13, such as a gathering of 13 persons or the 13th of the month. Friday the 13th continues to be regarded as specially unlucky. Even the word 'thirteen' is itself avoided by using such a circumlocution (or circumnumeration) as 'baker's dozen', 'long dozen', 'devil's dozen'. Societies are found in some countries that aim to challenge the bad luck of 13 direct, by deliberately choosing it on every possible occasion. In Toulouse, for instance, there is a society of 13 doctors who pointedly meet on the 13th of the month in room No. 13 of a local hotel, a practice they have been following regularly since 1854. But 13 can also be a

13—LUCKY FOR SOME!

Some people have found the number 13 to be lucky, not unlucky, especially in the world of sport.

The West German footballer Gerd Müller (b. 1945) regularly wore the number 13 when on the field. In 1970 he led the World Cup goalscorers with a total of 10 goals, and became European Footballer of the Year.

In athletics, the Bulgarian gymnast Maria Gigova (b. 1947) competed as No. 13 in the 1971 World Championship. She became Absolute World Champion that year, and the first woman to gain this title twice (having previously won it in 1969).

BISMARCK AND THE FIGURE 3

The famous German Chancellor, Otto von Bismarck (1815-98), led a life in which the figure 3 played an important role.

He served under 3 emperors, took part in 3 wars, signed 3 peace treaties, organized the union of 3 countries and meetings with 3 emperors, opposed 3 political parties, owned 3 estates and had 3 children.

His family crest bore the motto: *In trinitate fortitudo* ('In trinity, strength').

truly lucky number for some: for examples, see box on p. 53. Meanwhile, those who continue to dread unlucky 13 can be said to suffer from 'triskaidekaphobia', a term based on the Greek for '13-fear'.

17 The Greeks tried to avoid 17 whenever possible, and regarded it as an unfavourable number. Their basis for so doing lies in the fact that 17 stands between two 'good' numbers: 16, which is the square of 4, and 18, which is twice the cube of 2. Moreover, 16 is the only number that is the perimeter and area of the same square, so that the next number up from it will stand in unfavourable relation to it just as 13 does to favourable 12.

36 For the Greeks, 36 represented the highest oath as a 'swearing number'. This is because of its unique composition: 36 = 1 + 2 + 3 + 4 + 5 + 6 + 7 + 8; and also it equals (1 + 3 + 5 + 7) + (2 + 4 + 6 + 8), otherwise the sum of the odd and even numbers 1 to 8. As a result, the Greeks saw 36 as symbolic of the world, since it was formed out of the first 4 even and odd numbers, themselves the basis for many higher and more complex numbers.

40 The range of sayings, stories and superstitions centring on 40 in world literature is vast, and it has long been felt to be a 'top limit' number, an expression of completion. In the Bible, 40 is particularly well represented: Moses was 'in the mount' 40 days and 40 nights; Elijah was fed by ravens for 40 days; the rain of the great flood fell for 40 days, with another 40 passing before Noah opened the ark; the city of Nineveh had 40 days to repent; Jesus fasted for 40 days and 40 nights in the wilderness, and was seen 40 days after the Resurrection; while 40 men made a conspiracy against St Paul. In legend, meanwhile, St Swithin is associated with the 40 days of rain that are said to ensue if it rains on his feast day (15 July). In all such cases, and many others, 'forty' can be understood simply as 'many'. The fact that 40 is also halfway through an average person's life span, where it can well be a turning point, gives the number added significance. We say that life *begins* at 40, although we are already in mid-journey through it!

120 The Greek philosopher Philo of Alexandria attached much importance to the number 120. According to him, this was the number of years for which people lived after the Great Flood. Mathematically, it is certainly an interesting number. It is not only the sum of the first 15 numbers (1 + 2 + 3 + 4 . . . etc.), and a triangular number (one displayed with 15 units in the top row, 14 underneath, 13 underneath that, and so on), but the sum of 15, 25, 35 and 45, which are also triangular numbers. Moreover, it has 15 factors, 1, 2, 3, 4, 5, 6, 8, 10, 12, 15, 20, 24, 30, 40 and 60, which themselves sum to 240, twice 120. In this quality, Philo and his followers saw the essential duality of life — the life of the spirit and the life of the body.

666 The famous (or infamous) 'Number of the Beast' mentioned in the Bible (Revelation 13:18). There have been many attempts to interpret it, and it has featured in several literary works, notably Tolstoy's *War and Peace*. Many modern scholars are inclined to the view that the precise reference of the number was to the Emperor Nero. The Book of Revelation was probably written near the end of the 1st century AD, soon after the death of Nero (in AD 68), and if number-values are assigned to the letters of

BIBLICAL NUMBERS

The Bible is full of numbers, many of which had a mystic or magic significance (and which appear throughout this book in different guises). Others were historical or supposedly historical figures, used to denote a person's age, or to specify the number of soldiers in an army or an enemy force, or the number of members of a tribe.

The Bible even has a Book of Numbers, as the fourth book of the Old Testament. The book was so named with reference to the numbering of the tribes of Israel, described in Chapters 1-4. The Hebrew name for the Book of Numbers is more representative of its overall content, however. It is *Bemidbar*, 'In the Wilderness', referring to the Israelites as they wandered in the wilderness before entering the Promised Land.

THE NUMBER OF THE BEAST

In Leo Tolstoy's famous novel *War and Peace*, Pierre Bezukhov ponders on the significance of the number 666, the biblical 'Number of the Beast', when he is planning to kill Napoleon:

He felt that the position he was in could not go on for long, that a catastrophe was coming that would change the whole course of his life . . . One of his brother masons had revealed to Pierre the following prophecy relating to Napoleon, and taken from the Revelation of St John (13:17): 'Here is wisdom; let him that hath understanding, count the number of the beast; for it is the number of a man, and his number is six hundred threescore and six' . . . If the French alphabet is treated like the Hebrew system of enumeration, by which the first ten letters represent the units, and the next the tens, and so on, the letters have the following value:

a	b	c	d	e	f	g	h	i	k	l	m	n	o
1	2	3	4	5	6	7	8	9	10	20	30	40	50

p	q	r	s	t	u	v	w	x	y	z
60	70	80	90	100	110	120	130	140	150	160

Turning out the words *l'empereur Napoléon* into ciphers on this system, it happens that the sum of these numbers equals 666, and Napoleon is thereby seen to be the beast prophesied . . . This prophecy made a great impression on Pierre. He frequently asked himself what would put an end to the power of the beast, that is, Napoleon; and he tried by the same system of turning letters into figures, and reckoning them up to find an answer to this question. He wrote down as an answer, *l'empereur Alexandre? La nation russe?* He reckoned out the figures, but their sum was more or less than 666. Once he wrote down his own name 'Comte Pierre Bezukhov', but the sum of the figure was far from being right. He changed the spelling, putting *s* for *z*, added 'de', added the article 'le', and still could not obtain the desired result. Then it occurred to him that if the answer sought for were to be found in his name, his nationality ought surely to find a place in it too. He tried *Le russe Besuhof*, and adding up the figure made the sum 671. This was only five too much; the 5 was denoted by the letter 'e', the letter dropped in the article in the expression *l'empereur Napoléon*. Dropping the 'e' in a similar way, although of course incorrectly, Pierre obtained the answer he sought in *L'russe Besuhof*, the letters of which on that system added up to 666. This discovery greatly excited him . . .

(Leo Tolstoy, *War and Peace*, translated by Constance Garnett)

the two Hebrew words that mean 'Emperor Nero', these add up to 666. The fact that the apocalyptic number appears as 616 in some early versions of the text can similarly be justified, for if Nero's name is written in the Latin manner, without the final 'n', which had a value of 50, then the tally is accurate on this count also. But it is possible to juggle number-values with the letters of other names, too, to produce the original 'beast', and this has been done to make 666 fit Trajan, Caligula, Luther, Muhammad, Napoleon and many others.

Numerologists, on the other hand, are more inclined to total the three 6s in their accustomed manner to make 18, and to add these two digits in turn to make the magic number 9 (see above).

It has also been pointed out that 666 is just 1 short, three times, of the mystic number 777, perhaps a deliberate reference or device on the part of the author of Revelation (traditionally St John the Divine).

MAGIC SQUARES

Numbers and 'magic' combine more realistically in the **magic square**. A magic square is a grid of numbers from 1 upwards in sequence, with rows that add up to the same total, the magic constant, whether read across, down or diagonally. The square itself can be of any size from 3-by-3 upwards. (A 1-by-1 square would obviously consist trivially of a figure 1 in a single box, while a moment's trial will show that there can be no 2-by-2 square, where the numbers would have to be 1, 2, 3 and 4.)

The 3-by-3 or **order-3** square is probably the most familiar as it is unique. There is only one possible such square, with a magic constant of 15 for its added rows, columns and diagonals. Obviously, the figures themselves can be rearranged within the square by being rotated and reflected (as if in a mirror), but the basic disposition of the numbers 1 to 9 will always be:

6	1	8
7	5	3
2	9	4

This 'model' magic square dates back at least 2000 years to ancient China.

On progressing to an **order-4** square, however, there are no less than 880 different ways of arranging the numbers 1 to 16, and in the 16th and 17th centuries people often amused themselves seeing how many of the 880 they could find, much as people solve crosswords today. (A record of all 880 was first published in 1693.) Here is one such square (magic constant of 34):

16	3	2	13
5	10	11	8
9	6	7	12
4	15	14	1

Moving on to **order-5** squares, the number of possible squares is so great that it has only recently been calculated by computer. It is an astonishing 275,305,224 squares. The magic constant of each is 65, and one example is as follows:

8	22	5	6	24
7	14	15	10	19
25	9	13	17	1
23	16	11	12	3
2	4	21	20	18

THE GOLDEN NUMBER

The Golden Number is the number used to calculate the date of Easter. It can be any number from 1 to 19, the latter being the number of years in the Metonic cycle (named after Meton, the 4th-century BC Greek astronomer who discovered it), at the end of which the new moons fall on the same days of the year. The Book of Common Prayer gives the rule for finding the Golden Number: 'Add one to the Year of our Lord, and then divide by 19; the Remainder, if any, is the Golden Number; but if nothing remaineth, then 19 is the Golden Number'. So for 1990, the Golden Number is 14. The Book of Common Prayer provides tables to show that Easter Day this year falls on 15 April. The Golden Number is so called because in the Roman calendar it was marked in gold.

This particular square happens to contain another, order-3 square in its 9 innermost numbers (centring on the 13). This produces a magic constant of 39. Of course, it 'cheats', since it does not use the numbers from 1 to 9, as a proper 3-order square should. But if 8 is subtracted from each number in this square, we get our original order-3 square, only turned through 90°.

Order-4 squares can also be constructed with enjoyable additional features, and astute readers may have noticed that the example of the order-4 square given above is one of them. Not only do its rows, columns and diagonals add up to 34, but so do the 4 numbers in its 4 corners, as well as each of the 4 'inner' squares in the top left, top right, bottom left and bottom right of the square, plus the central square. Nor is that all, for every row has two adjacent numbers adding up to 15 *and* a pair of adjacent numbers that add up to 19. Magic indeed!

When it comes to squares greater than order-5 there are so many possible examples that they have not yet been calculated. All that can be said at present is that the number of possible magic squares for each order is as follows:

order-1 1
order-2 0
order-3 1
order-4 880
order-5 275,305,224
order-6 . . . ?

Finally, here is a special **order-7** magic square created by the 16th-century German mathematician Michael Stifel:

40	1	2	3	42	41	46
38	31	13	14	32	35	12
39	30	26	21	28	20	11
43	33	27	25	23	17	7
6	16	22	29	24	34	44
5	15	37	36	18	19	45
4	49	48	47	8	9	10

LUCK IN ODD NUMBERS?

The common belief that 'there is luck in odd numbers' is not supported by scientific or mathematical proof. At the same time, there is no doubt that, from the earliest times, odd numbers such as 3, 5, 7 and 9 *have* come to be regarded as specially significant or propitious. The saying is an old one, and echoes Virgil's words: *Numero deus impare gaudet*, 'A god delights in an odd number' (*Eclogues*).

SATAN BANISHED BY US PRESIDENT

When US President Ronald Reagan retired in 1989 after eight years in office, he and his wife Nancy moved from the White House to a private residence in the district of Bel-Air, Los Angeles. The address of their new home was 666 St Cloud Road, but in order to avoid this 'beastly' number, the Reagans successfully applied for it to be changed. They now live at No. 668.

This has a magic constant of 175. But in addition, its inner section is an order-5 square, using the numbers 13 to 37, and that in turn has an inner section that is an order-3 square, using the numbers 21 to 29.

Stifel included this magic square in his mathematical study *A Complete Arithmetic*, published in 1544, and this was the first work to include a scientific examination of the properties of magic squares.

6 NUMBERS AT WORK AND PLAY

PROBLEMS AND PUZZLES WITH NUMBERS

Numbers are not simply abstract things, any more than letters are. They are practical and meaningful, and can be employed and manipulated in many different ways, from an everyday shopping calculation to the solving of problems purely for the satisfaction of solving them, much as many people enjoy solving crosswords.

The art (or science) of using numbers and number concepts like this is known as **recreational mathematics**, although this phrase may seem a contradiction in terms. How can mathematics (associated still with the schoolroom and mental torment) be bracketed with recreation (associated with the schoolyard and sport)?

The answer is that the image of mathematics and 'figure-work' generally has become tarnished for older people by the memory of classroom discipline, with its formidable arithmetic, geometry, algebra and mathematics generally, all severe-seeming subjects studied for examination purposes.

But the truth of the matter is that we all wish to be as numerate (competent to deal with numbers) as we are literate (competent to deal with words). And if word-handling can be a pleasurable experience (conversation, reading, crosswords, etc.) so can number-handling. It is simply that the medium is different, or that the 'language' is different.

So this chapter is about number-handling. It will involve the reader more actively than most of the other chapters since he or she will be offered an intentionally wide range of situations to handle, from the purely verbal to the patently visual.

Since this book is not a mathematical textbook, the reader will not be laboriously guided through the various steps needed to reach the answers or solutions. It will be found that a combination of logic, patience and even 'trial and error' will do the trick, together with any mechanical aids such as a calculator that the solver wishes. At the same time, some of the more complex or possibly less familiar manipulations are explained in the solutions (on p. 64).

There are 10 numerical challenges in all, and the average reader should be able to solve at least half of these without any difficulty. Solve all of them, and you can regard yourself as numerate! (The tasks have all been specially devised or adapted for this book. In many cases, the reader will find it easy to copy the visual problems on to a separate sheet of paper, in order to avoid marking the book page.)

The tasks are not arranged in order of ascending difficulty, although the verbal items do precede the visual.

1 Mike was lucky enough to win £500 from his Premium Bond holding just before he entered college. He decided to bank £20 of it, and splash out the rest on things he wanted while in residence. These were a TV, a radio, a camera, and a new electric razor. The radio, camera and razor cost £125 together. The TV, camera and razor amounted to £420. The radio and the camera together were £100. How much were each of the four items individually?

2 The barber, Alan, usually has a chat when I am in his chair. Last week he mentioned that he had just had a birthday, and when I asked him how old he was, he said that he was exactly twice as old as the ages of his two daughters combined. He added that two years ago he was four times older than Sally, his elder daughter, and that four years ago he had been six times older than Jenny, his younger girl. As I sat there, I was able to work out the ages of all three from what Alan had said. How old were they?

3 The police dog handlers were holding a contest for their charges. The dogs, fine German shepherds, were named Copper, Victor, Umber, Nipper and Andy, and their handlers were Len, Geoff, Steve, Kevin and Chris. (I have not matched the handlers' names with those of the dogs.) The contest was quite straightforward, and the dogs had to carry out two tests: (a) find a hidden object, (b) climb and descend a step-ladder with a platform at the top. They were awarded 20 points maximum for each task, and, for the first test, Victor got 19, Copper got 16, and Nipper was given 18. Geoff's dog got the maximum score, but Kevin's charge got 15. For the second test, Umber came out on top, with a full 20 points. Chris's dog was close behind, with 19, while Andy and Nipper were 4 points behind *him*. Len's dog got 18, though. As he watched the others competing, Chris wished his dog had done as well as Victor, while Kevin envied the way that Umber's handler had trained him. So how many points did each dog score in total for the tests, and which handler trained which dog?

4 Uncle John often tried to catch out his clever young niece with number problems. He considerately waited until Gillian had finished her homework one evening, then when she came in to watch TV said, just before she switched on, 'Try this, Gillian'. He produced a smallish opaque carrier bag. 'I was in Woolworths today and I bought two tennis balls. They are both in the bag, and each of them could be either white or yellow.' 'Oh yes?' said Gillian, intrigued by any challenge to her powers of deduction. 'Now, see if you can tell me what colour they really are, without taking them out of the bag.' 'Without taking even *one* out?' 'That's right,' said Uncle John. But Gillian got the correct answer—and Uncle John gave her the balls to keep. If a 10-year-old can solve this logical problem, surely you can?

5 You can produce a total of 1000 by using eight 8s and only a plus sign, thus: 888 + 88 + 8 + 8 + 8 = 1000. Allow yourself a little more latitude, however, and by using only plus and multiplication signs, as well as any type of fraction (for example 88/88 to make 1, or 88/8 to make 11), see if you can produce the following totals with the given number of digits: (a) Eight 7s to make 700; (b) Ten 4s to make 500; (c) Ten 9s to make 1000; (d) Ten 6s to make 600; (e) Eight 2s to make 200.

6 A *cryptarithm* is a sum in which letters have been substituted for numbers (the digits 0 to 9, with different letters for different digits). See if you can solve the following cryptarithm, using a calculator if desired to check the letter-values. The sum can be worked mentally, however, without a calculator. (Clue to start: in the second multiplication row, D × A = D, so A must = 1, the only digit that will produce the same answer number as its multiplicator.)

$$\begin{array}{r} A\ B\ C \\ D\ E \\ \hline F\ E\ C \\ D\ E\ C\ \ \\ \hline H\ G\ B\ C \end{array}$$

7 Rearrange the ten even numbers in the figure below so that the sum total of the four numbers in each of the large triangles (left-hand, right-hand, lower and central) is 50.

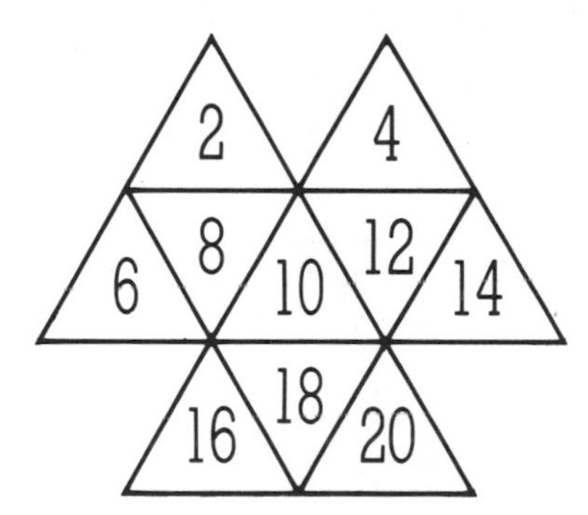

8 Write the numbers 1 to 20 in each of the circles in the figure in such a way that the sum of the five numbers in each of the five triangular points of the star is 50. This same total is also the one for the five numbers at the peaks of the points (on the edge of the outer circle).

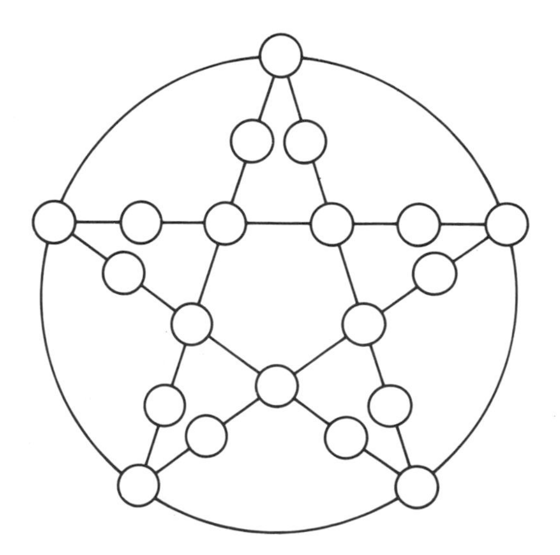

9 The 16 draughtsmen below are arranged in such a way that they form ten straight lines of four men (four horizontal lines, four vertical, and two diagonal). Add one more piece to the total, and rearrange the pattern completely *but keeping its symmetricity* in such a way that it shows eight straight lines of three men each.

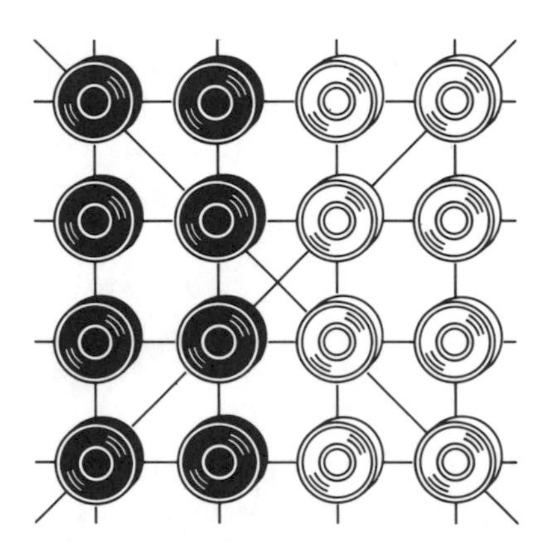

10 Reposition the barrels in the picture below by putting one barrel on top of another in such a way that you finish up with five groups of three, stacked on top of one another. Do this in only ten moves, one barrel at a time, with the barrel jumping over only *three* others to its new position. It can do this in either circular direction, and can of course leap over three barrels that have already been stacked. The numbers on the barrels are purely for identification purposes.

THE POCKET CALCULATOR

Pocket and larger calculators have transformed the everyday lives of many people, from schoolchildren solving mathematical problems to pensioners working out the weekly budget. An instantaneous touch of a button or two will display the required addition, subtraction, division, multiplication, percentage or square root, to mention only the most basic six functions offered by almost all calculators.

But as they can be put to work, so calculators can be made to play, or at least to entertain. Here are a few 'tricks' that can be offered to an appreciative viewer, such as a person who has recently acquired a calculator and, having mastered its basic functions, is looking for something more original.

1 Ask the owner to enter 98765432 and to divide this by 8. He or she will be surprised to find that the result is 12345679, with the digits in sequence except for the 8, which of course was 'used up' to divide the original number, also in sequence (but in reverse).

2 Ask the owner to enter 12345679 (or else to retain this number if the previous trick has been worked). Then ask him to name a number between 1 and 9. (If a child, his or her age will do well, if under 10.) Supposing he says 8. Tell him to multiply the displayed number (12345679) by 72. The result will be 88888888 (or nine 8s if the screen can display this many digits). What's the trick? The number you tell the owner to multiply by is always the product of 9 and the digit he or she chooses, in this case 8 ($\times$ 9 = 72). This works because 9 $\times$ 12345679 = 11111111, so that any number multiplied by the product of itself and 9 will give that number repeated in a row.

3 This trick can be effectively worked with the 'tricker' turning his back on the calculator owner, so that he is obviously not looking at the screen. Tell the owner to enter any three-figure number twice, such as 345345. Then start your patter, on lines such as the following: 'Hm—I hope you're not going to be unlucky. The number you have entered can be divided exactly by 13, without any remainder. See if I'm right: divide it by 13.' The owner does so. Behold—no remainder! You continue: 'Well, the next number looks luckier.

The number you have now can be exactly divided by 11, with no remainder. Divide it by 11.' The owner does so. Right again! Warming to your task, you go on: 'Right, now divide the number you have by 7. You'll find that that has no remainder, either.' Once again, true! Still with your back to him, play your trump card. Tell him to look closely at the number displayed. It will be the three-figure number that he entered twice in the first place. How does this work? Simply because if you multiply any three-figure number by 1001 it will obviously produce that number twice repeated. And because the factors of 1001 are 13, 11 and 7, dividing the twice-repeated three-figure number by those three numbers must reduce it to its original. If in doubt, try this yourself first, with any such number.

4 This trick take a little longer, but is good addition practice and produces a satisfyingly effective result. Before beginning, tell the calculator operator that you have written three numbers on a piece of paper and that he can work the calculator to produce these very figures by choosing any number he likes as a base for the sum. The numbers you actually write (on separate bits of paper, ideally) are 1, 6 and 8. Ask your companion to choose any three-figure number and write it down. Ask him then to write any further three-figure number he likes under the first, and to add them together. He can start to use his calculator at this point. He then adds the sum he has obtained to the second number (the one he chose) to produce a fourth number. He goes on like this, adding sum to previous total, and writing down the results as he goes, until he has a list of 20 numbers (including the first three-figure number that he chose). When he has done that, ask him to divide the 20th number by the previous figure, and to make a special note of the first three figures after the decimal point. These will almost always be, although not necessarily in the order you wrote them, the number you had predicted, consisting of 1, 6 and 8. He can even divide the 19th number by the 20th and get the same result.

What the operator has generated is basically a sequence of Fibonacci numbers (see p. 41), with each new number being the sum of the two immediately preceding it. By the time he has reached the 20th number, he has approached the so-called Golden Ratio, 1.618033 . . . , which is the ratio of AB to BC (or any corresponding ratios) in the pentagram, the symbol of health, as shown below.

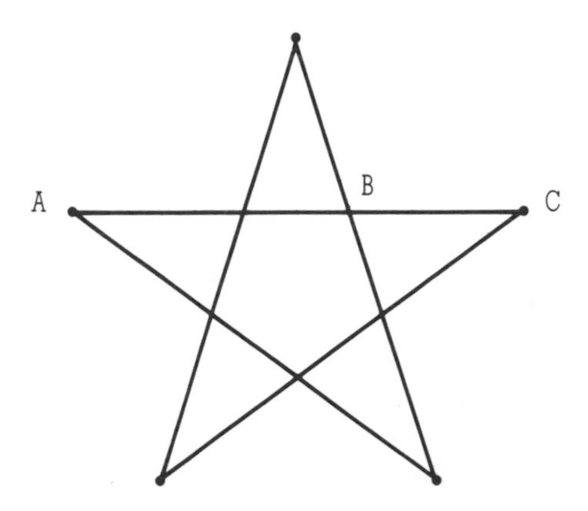

Mathematically, this is expressed as $\sqrt{5} + 1/2$, and was important to many ancient peoples, such as the Greeks and the Egyptians. The Golden Ratio was known as the Divine Proportion by Renaissance artists, who deliberately incorporated it in their works.

5 This final trick is different in nature. For some time now, calculator users have been aware of the fact that certain figures can be read as letters if the display is read upside down. One of the earliest manipulations of this feature involved posing a question about who won a certain war between the Arabs and the Israelis. The operator was given a simple calculation to work, which produced the result 71077345. Upside down, this reads as SHELLOIL. This sort of thing is adult whimsy, of course. The following is another example (and another sort of humour): What do politicians and go-go dancers have in common? Multiply 2417 by the number of signs of the zodiac, divide by the number of letters in 'Minister', then finally multiply by the number of letters in 'Margaret Thatcher'. Invert the screen to read the result. To round the figure out a bit, add 1.0956 to the displayed number, then subtract 0.1776. This particular trick is based on one devised by Martin Gardner ('my off-colour contribution', he calls it, 'to this useless pastime') in his book *Time Travel and Other Mathematical Bewilderments* (W. H. Freeman & Company, New York, 1988), where he has a chapter on tricks with calculators.

CODES

So far, we have mostly dealt with numbers in their conventional written or printed form, as Arabic or Roman figures, or as represented by a particular number of repeated symbols (such as ⁙ for '5').

But there are other ways of representing numbers in graphic form, and among them are different **codes**, some of which can represent figures and letters equally, while others are used just for figures.

The **Morse** code, for example, can transmit letters and numbers equally. The Morse symbols for the ten digits are:

1	· − − − −	6	− · · · ·
2	· · − − −	7	− − · · ·
3	· · · − −	8	− − − · ·
4	· · · · −	9	− − − − ·
5	· · · · ·	0	− − − − −

(Although it is also possible to use the same signals as for the letters A to J (or A to K, omitting J), preceded by a 'numerical' signal. The semaphore code does this, so that letters A to K double up as numerals 1 to 0.)

The Morse code, which owes its name to the American inventor Samuel Morse, was first used in 1838.

Much more recent, as another way of representing numbers both graphically and electronically, is the **barcode**, now found on almost all kinds of purchased goods in the form of a series of bars of different widths, with a row of figures underneath.

Barcodes first appeared on supermarket products in the early 1970s. What are they for, and what do they mean?

Basically, each 13-figure number uniquely identifies the particular product, in terms of its size, colour and packaging. The codes themselves are provided by an organization called (unsurprisingly enough) the Article Number Association (ANA), and are designed for use at a supermarket checkout, where the store is equipped with an EFTPOS (*e*lectronic *f*unds *t*ransfer at *p*oint *o*f *s*ale) system. When a customer makes a purchase, the checkout assistant, instead of entering the price into a cash register, reads the barcode by means of a laser scanner or other optical character reading device. The product's number is then transmitted to an electronic price file which instantly relays the product's price and description back to the cash register. In this way, the customer gets a till receipt which not only shows the product's price but also describes it. The process is speedy, and (theoretically) the checkout queue moves more rapidly, since the assistant has less to do. At the same time, information from the checkout passes to an in-store computer, where the collated data can greatly simplify the matter of stock checking or ordering.

But what do the 13 digits actually mean? They have no numerical significance in themselves, but purely provide a reference number. Observant shoppers will have noticed that the number usually begins with the two digits 50. This means that the product's number was issued in Britain by the ANA. The next five digits are allocated by the ANA to the company that markets the product. The next five after that are used by the company itself to give a unique number to each of its products, or each variation of a product. That makes 12 digits. So what about the 13th? The final digit is a computer check digit, so that if any mistake is made in composing the first 12 digits, the computer will instantly alert the point-of-sale operator that the number is incorrect.

Barcodes, incidentally, can be read from either end by the assistant's scanner, and their basic simplicity makes them easy to print. Even creased or stained barcodes are usually read by the machine at the first attempt without any difficulty or delay.

CODES AND CIPHERS

It goes without saying that numbers can be ideally employed for the construction or production of encoded messages.

An obvious way to use them is as a substitute for letters of the alphabet. This can be done on a simple substitution basis (A = 1, B = 2, C = 3, and so on, continuing as necessary), or, more neatly, by means of a grid, where figures can be read off as letters, or letters as figures. A classic example here is the so-called 'Greek square', which allows each letter of the alphabet to be enciphered by a two-digit number that represents the row (horizontally) and the column (vertically) in which the letter appears, thus:

	1	2	3	4	5
6	A	B	C	D	E
7	F	G	H	IJ	K
8	L	M	N	O	P
9	Q	R	S	T	U
0	V	W	X	Y	Z

In this, 48, for example, would represent O, and 49 produces T. The word GUINNESS, transmitted by this particular Greek square, transforms alphanumerically (or polyliterally) as 2759473838563939. But it would not take a cryptographer long to spot the repeated digits here for the two double letters (38 twice for N, 39 twice for S), and a more subtle code would not be hard to devise.

Some numerical codes have become famous in literature, such as the one in Edgar Allan Poe's short story 'The Gold Bug'. This utilized not simply the ten digits 1, 2, 3 . . . 0, but certain punctuation marks and other symbols, and the story's main character, William Legrand, finally succeeds in unravelling the following:

53‡‡†305))6*;4826)4‡.)4‡);806*;48†8¶60))85;;]
8*;:‡*8†83(88)5*†;46(;88*96*?;8)*‡(;485);5*†2
:*‡(;4956*2(5*—4)8¶8*;4069285);)6†8)4‡‡;1(‡
9;48081;8:8‡1;48†85;4)485†528806*81(‡9;48;(8
8;4(‡?34;48)4‡;161;:188;‡?;

The high frequency of certain figures and symbols, such as '8' and ';', suggest (correctly) that these represent letters of the alphabet that have a corresponding frequency. They are actually 'E' and 'T', respectively, and with his knowledge of letter frequencies, Legrand soon deciphers the whole message successfully, as: 'A GOOD GLASS IN THE BISHOP'S HOSTEL IN THE DEVIL'S SEAT—TWENTY-ONE DEGREES AND THIRTEEN MINUTES—NORTHEAST AND BY NORTH—MAIN BRANCH SEVENTH LIMB EAST SIDE—SHOOT FROM THE LEFT EYE OF THE DEATH'S HEAD—A BEELINE FROM THE TREE THROUGH THE SHOT FIFTY FEET OUT'.

Might not an all-numerical code have been more effective? It probably would, but, within the fictional bounds of the narrative, the message had to be in code, but at the same time in a code that could be deciphered without too much difficulty. An all-figure code might well have proved the cause of its own undoing.

The reader is invited to decipher the following numeric message. The answer is on p. 65. (It helps to be, or to know, a typist.)
56828212870937322153594313846973165603

CHRONOGRAMS

Roman numerals can be used for coded messages just as Arabic numerals can. The classic example of such punning usage is the **chronogram**. A chronogram is a word, phrase or sentence in which the letters that coincide with the Roman numerals (such as I, V, X, L, C, M) are specially selected so as to indicate an appropriate date. The pastime was popular in the 17th century, and in order to 'spell out' the required figure, the relevant letters were usually written or printed in capitals, although not necessarily in the right order.

The classic example, quoted in *Webster's Third New International Dictionary*, is the Latin motto of a medal struck by Gustavus Adolphus in 1632. This read: ChrIstVs DVX; ergo trIVMphVs ('Christ the leader; therefore a triumph'). It is a relatively easy task to pick out the Roman letters, CIVDVXIVMV, rearrange them in their correct

CRUCINUMERICS

Setters of cryptic crossword clues frequently resort to Roman numerals for their challenges. A common device is to combine a number with a noun in this way, so that '100 lights' produces CLAMPS (C LAMPS), and '1000 woes' leads to MILLS (M ILLS). A clever example of a complete clue like this is '500 + 500 = a vision'. The innocent solver tries to get '1000' somewhere in the answer. But the true solution is DREAM, otherwise D (500) + REAM (500 sheets of paper).

See if you can solve these cryptic 'crucinumerics':

1. 100 slopes can cause contractions of the muscles.
2. 50 snakes use these to climb.
3. Ten follow the queen into the extra accommodation.
4. Five narrow lanes lead to the dales.
5. 1001 and 10? That's a muddle!

Answers on p. 66.

order, MDCXVVVVII, and get (cheating a little, with VV = X) the year when the medal was struck, 1632.

Sometimes, a good chronogram can relate meaningfully to the date it denotes, like this one, referring to the St Bartholomew's Day massacre of 1572 in Paris: LVtetIa Mater natos sVos DeVoraVIt ('Mother Lutetia has devoured her own children'). Extract the Roman numerals: LVIMVDVVI; rearrange them: MDLVVVVII (with VV = X, as before), and you get the date 1572. (Lutetia was the Roman name of Paris.)

Chronograms do not necessarily have to be in Latin, of course, despite the Roman numerals, and an English chronogram that was composed to serve as a commemoration of the death of Queen Elizabeth runs: My Day Is Closed In Immortality. The Roman number here, with a minor adjustment of the letter order, is MDCIII, or 1603, the year of the Queen's death. An attractive feature of this sort of chronogram is the ability to begin each word with a capital letter, instead of singling them awkwardly out in the middle of a word.

In his *Oddities and Curiosities of Words and Literature*, originally published as *Gleanings for the Curious* in 1874, the American writer C C Bombaugh notes a remarkable work consisting of 100 hexameters, every one of which is a chronogram indicating the year 1634. This was the year when the Holy Roman Emperor Ferdinand II, to whom the work is addressed, was appointed Commander of the Imperial Armies and won a noted victory over the Swedes at Nördlingen. Bombaugh quotes the first and last lines of the work, which run respectively as follows:

> AngeLe CæLIVogi MICHaëL LUX UnICa CætUs.
> VersICULIs InCLUsa, fLUent In sæCULa CentUM.

(Here the Us serve as Vs. The first line refers to St Michael, to whom the work was dedicated.)

Readers may like to compose their own chronograms (in English, or other preferred language) to denote a year significant in their own lives. The full range of letters available is (in numerical order): I, V, X, L, C, D, M. It is true that X presents a problem, so possibly one answer is to choose a year that does not contain it, such as MCMLV (1955), although there are obviously more dates with X than without.

ANSWERS TO CHALLENGES (see p. 58)

1 The TV was £355, the radio £60, the camera £40 and the electric shaver £25.

2 Alan is 58, Sally 16, and Jenny 13.

3 The overall winner was Umber, with 40 points. His handler was Geoff. Next came Victor, scoring 37, and trained by Len. Third came Copper. He scored 35 points, and he was handled by Chris. Nipper was fourth, scoring 33, and trained by Steve, leaving Andy last, with 30 points. His handler was Kevin.

4 One ball is white and the other yellow. Whether Gillian was a good guesser or not is uncertain. But she may well have worked it out the logical way. If the bag had contained three balls, two white and one yellow, the chances of taking out a white ball would have been two out of three. But as there were not three balls, there must be some other chance. The chances of there being two white balls there are one out of four. The chances of one being white and one yellow are one out of two. The chances of both being yellow are, as for the two white, one out of four. Suppose you add a white ball to whatever is in the bag. The chances are as before: one in four that there will be three white, one in two that there will be two white and one yellow, and one in four that there will be two yellow and one white. The chances of picking out a white one are therefore two out of three (a quarter times one, plus a half times two-thirds, plus a quarter times one-third). Thus, with the white ball added, the bag must contain two white balls and one yellow, since any other combination would not give this two-thirds chance. So if there are (hypothetically) two white balls and one yellow, there must have been *one* white ball and one yellow originally.

5 (a) $77 \times 7 + 77 + 77 + 7 = 700$; (b) $444 + 44 + 44/4 + 4/4 = 500$; (c) $99 \times 9 + 99 + 9 + 99/99 = 1000$; (d) $66 \times 6 + 66 + 66 + 66 + 6 = 600$; (e) $22 \times 2 \times 2 \times 2 + 22 + 2 = 200$.

6 $125 \times 37 = 4625$. As mentioned, it can be immediately deduced that $A = 1$. It should then be noticed that $D \times C$ and $E \times C$ both end in C. The only multiple from 1 to 9 that will produce this is 5. So $C = 5$. D and E must be odd numbers, and as both partial products (FEC and DEC) have only three figures, neither D nor E can be 9. They must therefore be either 3 or 7, as 1 and 5 have already been allotted. In the first partial product, E × B is a two-figure number, while in the second partial product D × B gives just a single-figure number. E is therefore larger than D, so must be 7. D is therefore 3. Because D × B has only one figure, B must be 3 or less. The only two possibilities are 0 or 2 (as 1 has been allotted to A). It cannot be 0 because 7 × B is a two-figure number. So B is 2. Complete the multiplication sum, and you will find that F = 8, G = 6 and H = 4.

7

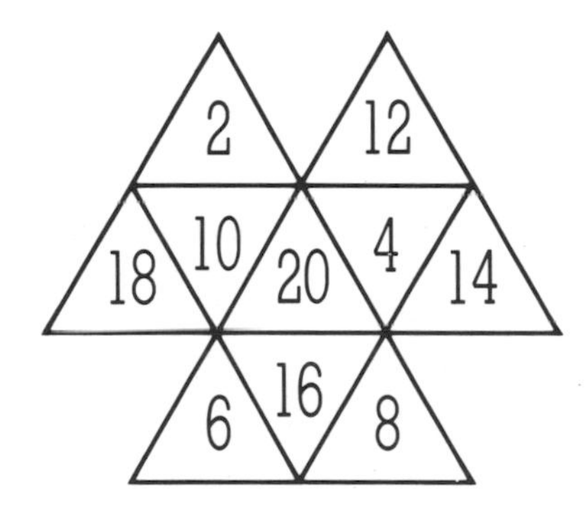

8

9

10 One solution: Barrel 5 jumps over 4, 3 and 2 to land on 1. Barrel 6 goes over 4, 3 and 2 to land on 1. There is the first stack of three barrels. Barrel 9 then leaps over 8, 7 and 4 to land on 3. Barrel 10 then goes over 8, 7 and 4 to land also on 3. There is the second stack of three. After this, barrel 8 goes over 11, 12 and 13 to land on 14. Barrel 7 follows it to land also on 14, making the third stack. Then barrel 4 jumps over the stack of three on 3 to land on 2, and 11 does likewise to land on 2, making the fourth stack of three. Finally, barrel 13 springs over the three barrels on 14 to land on 15, and 12 follows it, landing there also, making the fifth and final stack of three.

ANSWER TO NUMERIC CODED MESSAGE (see p. 63)

If you are familiar with the layout of a typewriter keyboard, you should soon spot that each digit represents one of the letters in the three rows of letters that, on most keyboards, are located under the numerals (as the top row), with each letter being in the same position, counting from the left, as its numeral. In other words the correspondences are as follows:

1	2	3	4	5	6	7	8	9	0
Q	W	E	R	T	Y	U	I	O	P
A	S	D	F	G	H	J	K	L	
Z	X	C	V	B	N	M			

The message thus reads:

5 6 8 2 8 2 1 2 8 7 0 9 3
T H I S I S A S I M P L E

7 3 2 2 1 5 3 5 9 4 3 1 3 8 4
M E S S A G E T O R E A D I F

6 9 7 3 1 6 5 6 0 3
Y O U C A N T Y P E

What makes the code slightly more complex than a simple substitution is that the digits 1 to 7 can have three possible corresponding letters. This shows up in a word like 494, where the 4 represents F the first time and R the second, to read 'FOR'. On the other hand, each letter will always be represented by the same digit, so that frequently recurring letters will have frequently recurring digits in the code. But, in turn, a high frequency of identical digits may be misleading, for the very reason that one digit can represent two or more letters! (Of the seven 3s, five represent E, one D, and one C.)

ANSWERS TO CRUCINUMERICS (see p. 63)

1 Cramps (C ramps)

2 Ladders (L adders)

3 Annex (Anne X)

4 Valleys (V alleys)

5 Mix (MI X)

7 LIVING BY NUMBERS

NUMBERS IN EVERYDAY LIFE

Not a single day goes by in our regular lives without one or more numbers being involved. It can be a date, an age, an address, a weight, a temperature, a size, a score, a result, a time, a grade, a standard, a rate, a quantity, or a dozen or more other things.

For some of the most common types of number, we can usually remember a whole range: people's ages, addresses, phone numbers, birthdays, and so on. In particular, our daily lives are more or less governed by the clock, with the time expressed numerically, and our days are measured by the calendar, with particular days and dates associated or set aside for particular events or activities. Not many of us can get by without a clock or watch on the one hand, and a calendar or diary on the other! And if we have a diary, or some other form of 'personal organizer', we may need to make note of our own personal numerical data: the sizes we take in clothes and footwear, the numbers of our various licences and accreditations, the birthdays of our friends and colleagues.

We cannot live without numbers, and for this reason we are interested to learn of deviations from the norm. But first we need to learn what the many norms are. In height, for instance, we know that someone 6 ft (1.8 m) or more is tall, and 5 ft (1.5 m) or under is small. In weight, we are aware that to be 14 st (89 kg) or more (for a man) is to be heavy or overweight, while to be under 7 st (44 kg) (for a woman) is to be thin, even anorexic. A man who takes size 12 shoes has big feet, while a woman who takes size 3 has tiny feet.

Then there are the 'legal limits': what a person may or may not do at a particular age. Or the progressive norms of life: what a child is expected to do at a particular age. Not only in our work, but in our play (or sport) we need to 'know the score', sometimes quite literally. We perhaps aim to be promoted to a higher grade or salary scale at work, and we will know what that is. When playing our favourite sport or game, we need to know what the scoring system is. A score of 40–0 in tennis looks like a near victory in a game for one player. But the same score in a football match would be laughable.

In an increasingly technological world, our knowledge of numerical facts and figures becomes more complex, but there are some areas and ranges of numbers that we encounter daily, beginning with birth and proceeding to the most mature and complex stages of our lives.

NAME THAT NUMBER

How many of the following everyday numerical facts can you supply? The answers are all given somewhere in this chapter. (See also p. 85.)

1. The normal temperature of the healthy human body.
2. The number to ring for 'Directory Enquiries' (in the UK outside London).
3. The year when the Age of Majority was reduced from 21 to 18 (in UK).
4. The number of years for a steel wedding.
5. The smallest grade of egg.
6. The number of the London–Basingstoke motorway.
7. The number of the first car to be registered.
8. The Dewey Decimal hundred-number for books on art in a public library.
9. The speed (in mph) of a wind force 6.
10. The alcohol strength (as a percentage) of gin.
11. The bra size for a woman with a rib cage measurement of 29 in.
12. The centimetre equivalent for a man's size 15½ shirt.
13. The first two digits of a Paris postcode.
14. The dialling code for Glasgow.
15. The usual number of pieces in a set of dominoes.
16. The points value of the brown ball in snooker.
17. Marilyn Monroe's 'vital statistics'.
18. The age to which Sir Winston Churchill lived.
19. The number of sheets of paper in a ream.
20. The life expectation of a woman at birth (in the UK).
21. The number of figures in a Soviet postcode.
22. The temperature that corresponds to Gas Mark 6.
23. The number of tablespoons used to measure a cupful.
24. The highest shoe size number (but not the largest size) for children.
25. The height of an average female at age 17.

SHIPSHAPE

The Cunard liner *Queen Elizabeth 2*, named after Queen Elizabeth II, and better known as the *QE2*, was launched in 1967 and entered service two years later. She replaced an earlier Cunard liner, the *Queen Elizabeth* (named after Queen Elizabeth, The Queen Mother), built in 1940, but withdrawn from service in 1970. This earlier ship's own predecessor was the *Queen Mary* (named after the wife of George V). The third liner in the series was provisionally designated the *Q3*, but was never built. The *QE2* was thus provisionally designated the *Q4* before she officially became the *QE2*. QED!

LEGAL MILESTONES

age (years)

5 Compulsory schooling begins
8 Age of Reason (*Doli Capax*), age of first possible offence
11 State secondary education begins
12 Mormon males can become deacons; can see category '12' film
13 Bar Mitzvah (Jewish boy's age of religious responsibility)
14 Age of Discretion (can be legally prosecuted for offence)
15 Can see category '15' film
16 Age of Consent (can legally marry); minimum school leaving age
17 Can drive a motor cycle or car
18 Adulthood begins (Majority reached)
20 Mormon males can become elders (in Melchisedek priesthood)
21 Former coming of age (to 1970); can become MP; consenting age for homosexuals; can adopt a child
22 Roman Catholic males can become deacons
23 Church of England males can become deacons
24 Roman Catholic and Anglican males can become priests
30 Roman Catholic and Anglican males can become bishops
55 Compulsory retirement age in armed forces
60 Age of retirement for women
65 Age of retirement for men; maximum age for jury service
70 Driving licence must be renewed
75 Roman Catholic bishops must retire
100 The Queen's Message (formerly Telegram) of congratulations

FROM AGE TO AGE

The age to which people live is always interesting, however long, however tragically short. Here are the ages to which some famous people lived (with year of death):

12 Edward V (1483)*
13 St Agnes (*c.* 304)*
14 St Pancras (*c.* 304)*
15 Edward VI (1553)
16 Lady Jane Grey (1554)*
17 Grand Duchess Anastasia (1918)*
18 Tutankhamun (*c.* 1340 BC)
19 Joan of Arc (1431)*
20 Catherine Howard (1542)*
21 Pocahontas (1617)
22 Buddy Holly (1959)
23 Edmund Ironside (1016)*
24 St Theresa of Lisieux (1897)
25 John Keats (1821)
26 Fatima (632)
27 Rupert Brooke (1915)
28 Caligula (AD 41)*
29 P B Shelley (1822)
30 Emily Brontë (1848)
31 Cesare Borgia (1507)*
32 Alexander the Great (323 BC)
33 Dick Turpin (1739)*
34 Henry V (1422)
35 W A Mozart (1791)
36 Lord Byron (1824)
37 Marie-Antoinette (1793)*
38 Louis XVI (1793)*
39 Cleopatra (30 BC)
40 Edward IV (1483)
41 Jane Austen (1817)
42 Robert F. Kennedy (1968)*
43 Edward II (1327)*
44 St Francis (1226)
45 Rasputin (1916)*
46 John F. Kennedy (1963)*
47 Attila the Hun (453)
48 Charles I (1649)*
49 Queen Anne (1714)
50 Alfred the Great (899)
51 Honoré de Balzac (1850)
52 William Shakespeare (1616)
53 Princess Grace of Monaco (1982)
54 Charles II (1685)
55 Henry VIII (1547)
56 George VI (1952)
57 Paganini (1840)

58 Charles Dickens (1870)
59 Oliver Cromwell (1658)
60 Calvin Coolidge (1933)
61 Walter Scott (1832)
62 Edward the Confessor (1066)
63 August Strindberg (1912)
64 Lyndon B. Johnson (1973)
65 Lewis Carroll (1898)
66 Archbishop Cranmer (1556)*
67 Catherine the Great (1796)
68 Edward VII (1910)
69 Elizabeth I (1603)
70 Hans Christian Andersen (1875)
71 Tennessee Williams (1983)
72 Confucius (479 BC)
73 Charles Darwin (1882)
74 Mark Twain (1910)
75 Archimedes (212 BC)*
76 T S Eliot (1965)
77 Le Corbusier (1965)
78 Dwight D. Eisenhower (1969)
79 Charles de Gaulle (1970)
80 William Wordsworth (1850)
81 Queen Victoria (1901)
82 Leo Tolstoy (1910)
83 Voltaire (1778)
84 Thomas A. Edison (1931)
85 Queen Mary (1953)
86 Frans Hals (1666)
87 Emperor Hirohito (1989)
88 Harry S. Truman (1972)
89 Arturo Toscanini (1957)
90 Winston Churchill (1965)
91 Somerset Maugham (1965)
92 Knut Hamsun (1952)
93 P G Wodehouse (1975)
94 George Bernard Shaw (1950)
95 Marshal Pétain (1951)
96 Pablo Casals (1972)
97 Bertrand Russell (1970)
98 General Weygand (1965)
99 Titian (1576)
100 Margaret Murray (1963)
101 Grandma Moses (1961)

* murdered, executed or martyred; others who died before their 'threescore years and ten' will have suffered a fatal illness or accident, committed suicide, or simply died young.

THE GRAND CLIMACTERIC

It was formerly believed (and still is by some) that every seventh year in a person's life (7, 14, 21, etc.) was critical, or climacteric, when some important change in health or fortune occurred. The odd multiples (7, 21, 27, 35, 49, 63, 77, 81) were (are) believed to be specially critical. But most critical of all is 63 (= 7 X 9), called the *Grand Climacteric*. Formerly, many people did not succeed in living past this age. But if they did, an alternative Grand Climacteric was held to arrive at age 81 (9 X 9). Either way, it is significant that 63 is the age now being considered in Britain as a retirement age for both men and women.

THREESCORE YEARS AND HOW MUCH?

Today, the average healthy adult in the western world lives a good deal longer than the traditional biblical 'threescore years and ten', or 70.

The number of years a person can expect to live on reaching a particular age is his or her **expectation of life**. For men and women in England and Wales the respective life tables in 1983–85, to the nearest completed year, were as follows:

at age	*males*	*females*
0	72	78
5	68	73
10	63	68
15	58	63
20	53	59
25	48	54
30	43	49
35	39	44
40	34	39
45	29	34
50	25	30
55	20	25
60	17	21
65	13	17
70	10	14
75	8	10
80	6	8
85	4	5

LOVE AND MARRIAGE . . .

Although any person's life can be measured by his or her age, a married person's life is marked not only by a birthday but by a **wedding anniversary**. Traditionally, presents are given on particular anniversaries according to a special material or stone. They become increasingly valuable and precious, as the ideal marriage itself should, and are:

anniversary	*name*	*anniversary*	*name*
1	paper	14	ivory
2	cotton	15	crystal
3	leather	20	china
4	flowers (fruit)	25	silver
5	wood	30	pearl
6	iron (sugar)	35	coral
7	wool	40	ruby
8	bronze	45	sapphire
9	pottery	50	golden
10	tin	55	emerald
11	steel	60	diamond
12	silk (fine linen)	70	platinum
13	lace		

Of these, the key anniversaries are the silver, ruby, golden and diamond.

SIZING UP—OR DOWN

Apart from ages and anniversaries, there are other figures and numbers that relate directly to every individual. Most of us, for example, know how tall we are and how much we weigh. Women, too, will know their 'vital statistics' (bust, waist and hip measurements), whether for purposes of fitness and slimming or simply to buy the right size of clothing.

At the back of most of our minds there is an ideal measurement to which we should like to conform. What this measurement is will vary between the sexes, and will often depend on another personal parameter. A tall person can expect to weigh more than a small person, for instance, and women usually have lower 'averages' than men.

For many people, an ideal height and weight is the one recorded as average for a young man or woman on reaching adulthood, conventionally at age 17. At this age, the average male is 5 ft 10 in (1.7 m) tall and weighs 10 st 11 lb (151 lb) (68 kg), while the average female is 5 ft 4 in (1.6 m) tall and weighs 8 st 13 lb (125 lb) (57 kg). At the age of 48, the international pop star Cliff Richard, 5 ft 11 in (1.8 m) tall, weighed 10 st 9 lb (149 lb) (67 kg) and was satisfied with this. For an ideal 'Miss World' figure, on the other hand, most women would perhaps regard measurements of 34-26-36 as reasonable, although statistically 'Ms Average Slimmer' is likely to be nearer 38-30-40. Marilyn Monroe was 36-23-36, a figure which, with its matching bust and hip measurements, is frequently regarded as perfection.

Women's clothing sizes are conventionally given in a range of even numbers ascending in 2s from 8 (regarded as Small, or S) through 10, 12 and 14 (regarded as Medium, or M) to 16 (Large, or L) and beyond (referred to as Extra Large or XL). Each of these numbers corresponds to a fairly standard bust, waist and hip size, expressed in inches, as above. But the precise measurements can vary, especially for the waist, so that many mail order and fashion catalogues, for example, give a table for purposes of reference. The following is an example:

size	*bust*		*waist*		*hips*	
	(in)	(cm)	(in)	(cm)	(in)	(cm)
8	30	(76)	20	(51)	32	(81)
10	32	(81)	22	(56)	34	(86)
12	34	(86)	24	(61)	36	(91)
14	36	(91)	26	(66)	38	(97)
16	38	(97)	28	(71)	40	(102)
18	40	(102)	30	(76)	42	(107)
20	42	(107)	32	(81)	44	(112)
22	44	(112)	34	(86)	46	(117)
24	46	(117)	36	(91)	48	(122)
26	48	(122)	38	(97)	50	(127)
28	50	(127)	40	(102)	52	(132)
30	52	(132)	42	(107)	54	(137)
32	54	(137)	44	(112)	56	(142)

This table cannot be used, however, when a woman wishes to buy a **bra**. She then needs to know the measurement (in inches) of her rib cage (that is, the circumference of the body immediately below the bust) and to make two simple calculations based on this to obtain both the bra size and the cup size. To get the bra size, the normal procedure is to add 5 (inches) if the rib cage measurement is an odd number, and 4 if it is an even number. Thus a rib cage measurement of 29 will mean a bra size of 34 (29 + 5), while one of 28 will mean a bra size of 32 (28 + 4). The cup size, which (in order to avoid confusion) is expressed by letter, not number, is derived from the bust measurement itself, that is, at the fullest part. If this is the same as the bra size in inches, the cup size is A. If there is a difference of 1 between the actual measurement and the bra size, the cup size is B, and so on. This can be shown in tabular form, as follows:

	cup size	
(Small)	A	bust = bra size
(Medium)	B	bust = bra + 1
(Large)	C	bust = bra + 2
(Full)	D	bust = bra + 3
	DD	bust = bra + 4
	E	bust = bra + 5

Men buying clothes do not need to make such calculations. Nor do they have a distinct size number. All they need to know are their actual measurements: typically those of the circumference of the chest and waist (in inches). For trousers, they need to know their waist size and additionally the length of the leg. **Jacket** sizes are usually regarded as follows:

	Small	Medium	Large	Extra Large
size (in):	34 36	38 40	42 44	46 48 50

The waist size is usually about 5in (12cm) less than the chest measurement (jacket size), so that a size 38 jacket will need trousers size 33, for example. The leg length varies little, and conventionally is 32.

Children's clothes are often sized by age, especially for babies and younger children. The mother will thus look for '3' on a label for a 3-year-old, and '9' for a 9-year-old. But children's clothes are also measured by height and chest size (in inches), with both these parameters conventionally assigned to a particular age. A typical table runs as follows:

age	*height* (in)	(cm)	*chest* (in)	(cm)
3 months	25	(62)	17	(45)
6	27	(68)	18	(46)
9	29	(74)	19	(48)
12	32	(80)	20	(51)
18	34	(86)	20	(51)
2 years	36	(92)	21	(53)
3	38	(98)	21	(53)
4	41	(104)	22	(56)
5	43	(110)	23	(58)
6	45	(116)	24	(61)
7	48	(122)	25	(63)
8	50	(128)	26	(66)
9	53	(134)	27	(68)
10	55	(140)	28	(71)
11	57	(145)	29	(74)
12	60	(152)	30	(76)
13	62	(158)	32	(81)
14	64	(164)	34	(86)
15	66	(168)	36	(92)
16	66	(168)	36	(92)

Generally, however, boys' clothes use sizes based on height only from about the age of 7, as from this age on boys and girls develop at different rates. (Boys have larger chests and waists and smaller hips than girls.) And as children vary in their rate of growth individually, it follows that a size 5, for example, would be the right one for a 4-year-old who was 43in (110cm) tall.

Although body measurements, and many sizes, are still usually given in inches, many

HOW MANY CAN PLAY?

A concise guide to the standard number of players in a side for the most familiar sports. Where a side is usually named after its number, as in cricket, the name is added as a *word*.

American football 11
Association football (soccer) 11
baseball 9
basketball 5 or 6
bowls 1 to 4
cricket 11 (*eleven*)
croquet 1 or 2
curling 4
fives 1 or 2
Gaelic football 15
golf 1 or 2
handball 11
hockey 11
hurling 15
ice hockey 6
korfball 12
lacrosse 12
netball 7
polo 4
rackets 1 or 2
Rugby League 13
Rugby Union 15 (*fifteen*)
shinty 12
softball 10
squash 1 or 2
stoolball 11
table tennis 1 or 2
tennis 1 or 2
volleyball 6
waterpolo 7

BINGO!

The game of bingo (lotto, house, tombola), where random numbers are marked on cards, has evolved its own picturesque substitute calls to replace the prosaic numbers. They include the following (with the word 'number' usually called before numbers 2 to 10 or 11):

1 number one, just begun
2 dirty old Jew (popular formerly, now avoided)
3 cup of tea (or some popular name, e.g. Dixie Lee)
4 door to door
5 Jack's alive
6 chopsticks
7 hope in heaven (or some other word with 'heaven')
8 garden gate
9 doctor's orders ('No. 9 pill', formerly popular)
10 Big Ben, cock and hen, Downing Street
11 legs eleven (two 1s look like two legs)
13 unlucky for some
17 never been kissed (saying, 'sweet seventeen . . .')
21 key of the door (former age of majority)
22 dinky doo, all the twos*
26 bed and breakfast (formerly costing 2/6d)
39 all the steps (novel title, *The Thirty-Nine Steps*)
50 halfway house
66 clickety-click (random rhyme)
76 she was worth it (marriage certificate once cost 7/6d)
100 top of the house
* so also for 33, 'all the threes', 44, 'all the fours', etc.

manufacturers also quote sizes in centimetres. This means, for instance, that the size of a man's **shirt**, which is the circumference of his neck, will be 15½ (a 'medium' size) in inches but 39 in centimetres.

Shoes involve one of three possible sizing systems, the traditional 'English', the 'American', and the 'Continental'.

The 'English' system begins (theoretically) at 0, and runs up to 13 before starting all over again at 1 to climb again up to 13 (men's large size) or even occasionally 14 or 15. Size 0 begins at a length of 4 inches, with each succeeding size being one-third of an inch larger than the preceding one. This means that size 3 equates to a foot length of 5 inches, size 6 to 6 inches, size 9 to 7 inches, size 12 to 8 inches, big size 2 to 9 inches, big size 5 to 10 inches, big size 8 to 11 inches, and big (men's) size 11 to 12 inches, otherwise one foot, which is what one would expect.

Of these sizes, 1 to 13 and big 1 to 5 are normally regarded as children's sizes, while women's sizes generally run from 3 to 8 and men's from 6 to 11, although there are obviously some adults who will take sizes outside these ranges.

The 'American' sizing system is based on the 'English', and similarly runs up to 13 before beginning again at 1. However, although the sizes are still one-third of an inch apart, the starting point (notional size 0) is not exactly 4 inches, as in the 'English' system, but 3¹¹⁄₁₂ inches, a fraction shorter. This means that all 'American' sizes are slightly smaller than the 'English'.

The 'Continental' system, sometimes called the 'French', is not based on inches, of course, but on metric measurements. It starts at children's size 15, which corresponds exactly to the 'American' (but not the 'English') size 0, even though the latter is not a metric scale. It then progresses with intervals of two-thirds of a centimetre (*not* a whole centimetre, which might be more logical), which means that there are approximately 5 'French' sizes to 4 'American'. It runs up to size 45, which is just larger than an 'English' size 10 but slightly smaller than an 'American' size 11.

Unlike, the 'English' and 'American' sizes, too, the 'Continental' system has no half-sizes, as the smaller unit of measurement makes this unnecessary. (Half-sizes were first introduced in America in 1887, and were subsequently adopted by the British.)

A general trend towards metrication in recent years, boosted in this particular field of measurement by the complexity and illogicality of the existing shoe sizes, has resulted in much talk about a *genuinely* metric scale of shoe sizes, with the centimetre as the unit. The new system was designated **Europoint**, although this name was subsequently changed to **Mondopoint** (from French *monde*, 'world' and *point*, 'point'). But although some countries of the world have now adopted Mondopoint (South Africa was the first, by 1980), neither Britain nor the United States has yet managed to make the switch.

How does Mondopoint work? It bases its actual sizes on millimetres, with either 5mm (for fashion footwear) or 7½mm (for basic shoes) between sizes. This gives a correspondence of Mondopoint size 170 to 'English' children's size 10, and Mondopoint size 280 to adult 'English' size 10, both these on the scale that uses 5mm steps. (With steps of 7½mm, Mondopoint size 172 corresponds to 'English' children's size 10, and 277 to adult size 10.)

So much for shoe **lengths**. For shoe **widths**, a lettering system is used for women's shoes in both 'English' and 'American' scales, but a numbering system for men's shoes in 'English' sizes. American women have the luxury of no less than 15 different widths, from AAAAAA (the narrowest) to EEEEEE (the widest). Normally, however, widths run from A to E (or H for American men), while 'English' men's shoe widths run from 1 (the narrowest) to 8 (the widest).

Like half-sizes, shoe widths were an American invention.

HEALTH AND WHOLESOMENESS

For our bodily health and diet, further numbering systems are usually familiar.

For measuring **body temperature**, either Fahrenheit or Centigrade can be used, with some thermometers marked dually to register either. The equivalents for the body range are:

	Fahrenheit (degrees)	*Centigrade* (degrees)
	95	35
HEALTHY	96.8	36
AVERAGE→	98.6	37
	100.4	38
	102.2	39
	104	40

MAN'S NUMERICAL SYMMETRY

The physical body of an adult human being is numerically almost symmetrical, with a vertical line dividing it into halves, pairs or binary multiples.

We can thus be seen externally to have one head and one body, but two arms (and hands) and two legs (and feet), themselves having 10 (5 X 2) fingers and 10 toes respectively.

The head itself has one nose and one mouth, joined by a single neck to the body, with its one set of sex organs.

But the head also has two eyes and two ears, while the nose has two nostrils and the mouth has two jaws and two lips.

The female body, too, has two breasts, while the male has two testes, corresponding to the female's two ovaries.

Internally, the body has one heart and one liver, but two lungs and two kidneys. It also has 24 (2 X 12) ribs.

Finally (although not exhaustively), the mouth contains 20 (2 X 10) primary or deciduous teeth in early childhood, but 32 (2 X 16) teeth in adulthood.

It is hardly surprising that two of our commonest reckoning systems are the binary and the decimal!

Obviously, higher and lower extremes are sometimes encountered, but this range will fit most household needs. (For a more general comparative temperature chart, see p. 74.)

Those who need to watch their diet, on specific medical or general health grounds, will be calorie-conscious, and know, for example, that a boiled egg contains 90 calories, but a quarter-pound of sausages 400. Of more general interest is the chart which shows how many calories a day are needed by people at different ages and stages of their lives. For a young, healthy adult, the amount can vary depending on the degree of activity, and a pregnant woman, for example, will need more calories than one who is not pregnant (while breast-feeding needs a higher total still). The sex of a person comes into the reckoning, too. But a typical general calorie table runs as follows:

Age	*Male*	*Female*
0–1	800	800
1–2	1200	1200
2–3	1400	1400
3–5	1600	1600
5–7	1800	1800
7–9	2100	2100
9–12	2500	2300
12–15	2800	2300
16–18	3000	2300
18–35	2700 (normal)	2200 (normal)
	3600 (very active)	2500 (very active)
35–65	2600	2200
65–75	2300	2100
75+	2100	1900

A **calorie** is defined as the amount of heat needed to warm a kilogram of water by 1°C.

EGG GRADES

A diet may or may not involve **eggs**. If it does, eggs can be purchased according to a specific range of grades, from 1 (the largest) to 7 (the smallest). Eggs are graded by weight, and the standard weights for the grades are as follows (in grams):

1 70
2 65 to 70
3 60 to 65
4 55 to 60
5 50 to 55
6 45 to 50
7 under 45

WHAT'S IN A DRINK?

When shopping for **drink**—alcoholic drink, that is—we are usually aware of the percentage figure printed on the bottle label or can. This is the percentage of alcohol by volume (not to be confused with the 'proof' level), and denotes the strength of alcohol in the particular product. The higher the percentage, the stronger the drink, with figures like these as typical:

3–7% beer, lager
8–12% wine
14.5% vermouth
17% fortified wines (sherry, madeira, etc.)
25% some liqueurs
40% spirits (gin, brandy, whisky, vodka, rum)

The alcohol content of a type of drink can vary, of course, but (at least in Britain) spirits are not found less than 37.5%.

HOME NUMBERS

Apart from the obvious numbers of address and telephone various numbering systems are found in the home.

In the **kitchen**, for example, there are not simply standard measures for cooking purposes but a special scale of figures for use in ovens.

As well as knowing that 1 level tablespoon of granulated sugar will be approximately 1 ounce, for example, or that a similar tablespoon of flour will equal 2 ounces, the average cook will need to utilize equivalents such as these, in addition to the quantities already mentioned above.

1 teaspoon	= 5 ml (millilitres)
2 teaspoons	= 1 dessertspoon
2 dessertspoons	= 1 tablespoon
16 tablespoons	= 1 cup
2 cups	= 1 pint

For the **gas cooker**, the so-called 'gas mark' ('Regulo') numbers are still widely used to indicate oven temperature. Their approximate equivalents are as follows:

-	240°F	very cool
½	265°F	
1	290°F	cool
2	310°F	
3	335°F	warm
4	355°F	moderate
5	375°F	fairly hot
6	400°F	
7	425°F	hot
8	450°F	very hot
9	470°F	

Temperatures generally will matter both at home and at work, indoors and out, and an awareness of the Fahrenheit–Celsius (Centigrade) correspondences is usually helpful. For the Celsius '10s', the equivalents are as follows:

°C	°F
100	212—water boils
90	194
80	176
70	158
60	140
50	122
40	104
30	86—a hot summer's day
20	68
10	50—a mild winter's day
0	32—water freezes
−10	14—a cold winter's night
−20	−4
−30	−22
−40	−40

(See also the medical temperature figures on p. 73.)

BROADCASTING BY NUMBERS

British radio and television channels have been steadily clocking up the numbers since the BBC was founded in 1922.

The first numbered radio channel was the Third Programme (see p. 100), which was introduced in 1946 and so designated by contrast with the existing Home Service and Light Programme. In 1967 it was renamed Radio 3, and that same year the Home Service was redesignated Radio 4, the Light Programme was renamed Radio 2, and a new service of pop music was introduced as Radio 1. In 1990 a fifth channel, Radio 5, will begin transmission.

The main BBC television service was launched in 1936, and in 1964 was redesignated BBC1, when a new channel, BBC2, began transmission. The main ITV (independent) channel had all along been regarded as a third service, but has so far not been officially named as such. In 1982, however, a further independent TV service, Channel 4, was introduced, and another programme, Channel 5, is due to begin broadcasting in 1993.

With the advent of satellite TV in 1989, and the onset of the biggest upheaval in the history of British radio and TV broadcasting, many more numbered services are likely to be introduced.

Radio and television also play an important role in the home. For an outline of the different numbered channels and stations, see box, p. 74.

SIGNIFICANT NUMBERS

If you live or work somewhere, you need an **address**, and nine times out of ten it will contain a number, if only in the postcode.

The most usual address number is that of the street where the premises are located. Many streets are traditionally numbered with odd numbers on one side of the road and even on the other, meaning that adjacent houses can have an address differing by two unit numbers (for example, 145 will be next to 147). And even if houses do not extend equally on both sides of the street, the odds or evens will continue as appropriate, and may not necessarily have a 'pair' on the other side.

Clearly, the numbering of houses to indicate a particular address is a logical and simple idea, and the first house numbers are believed to have originated in Paris in 1463, when dwellings on the Pont Notre-Dame were numbered.

London houses had begun to acquire numbers by the early 18th century, and *A New View of London*, published in 1708, records that the houses in Prescot Lane, Whitechapel, had been numbered by their inhabitants, who were refugees from Europe.

In 1765 an Act of Parliament was introduced which decreed that all London houses should be numbered, except those that had names, and the practice of numbering individual buildings is one still encouraged today by the Post Office, although it is recognized that there are some houses that remain with names only, not numbers, especially in rural areas.

The present **postcodes** developed out of the former postal districts, notably those assigned to areas of London in 1859. In their present form, they were introduced gradually, and experimentally at first, from 1944. By 1968, postcodes had been allocated to 75 locations in Britain, and the following year they were extended to cover all Britain's 20 million addresses.

In Britain, unlike many other countries, postcodes comprise a combination of letters and numbers. A typical postcode consists of four elements divided into two halves, comprising respectively the letter code of the area head office, the number of the district in that area, the number of the postman's delivery area, and letters indicating the location of the address.

For example, the postcode GU31 4LN indicates the area head post office of Guildford, Surrey (GU), the district of Petersfield, Hampshire, together with an area round this town (31), the street named The Causeway (4) and the section of houses on this street numbered 125 to 209 (LN).

Countries such as the USA, France and the USSR have a postcode (in the US known as a 'zipcode') that consists simply of figures, without any letters.

The US **zipcode** (with the first three letters standing for 'zone improvement plan') was introduced in 1963. The first three figures of the number identify the section of the country to which the mail item is to be sent, with the last two digits indicating the specific post office or zone of the addressee. Regular mail users soon become familiar with the particular three-figure number borne by individual places, such as 100 for New York, 021 for Boston, Massachusetts, and 606 for Chicago.

France also has a five-figure postcode, with the first or first two figures indicating the postal area. Paris, for example, is 75, Lyon and the Rhône *département* is 69, and Nice and the Alpes-Maritimes *département*, is 6.

Elsewhere, South Africa has a four-figure code, the USSR a six-figure one, but Canada a letter-and-number combination like the UK, but always with single letters and numbers alternating. (The postcode for the Canada Publishing Corporation, Ontario, for example, is M1S 3C7.)

Undoubtedly, alphanumerical codes like this are harder to remember and more cumbersome to write and type, but they can be more accurate (and meaningful) than purely numerical codes.

Telephone numbers were introduced in Britain in 1879, and until comparatively recently consisted of an exchange name followed by a particular telephone's number. Early or isolated telephone subscribers were frequently allocated low numbers, and some telephone numbers in single figures existed well into the 1930s. Thus the number of the Grand Hotel, Sheringham, Norfolk at this time was Sheringham 2, while the Hindhead Beacon Hotel, Haslemere, Surrey was Hindhead 7.

Today, although exchanges still have names,

for practical purposes they have been superseded by numerical dialling codes. The numbers overall are not only longer but may have changed altogether over the years.

Here are the telephone numbers of some Oxford hotels as they were in the 1920s and as now:

	1922	*1988*
Randolph Hotel	Oxford 290	0865-247481
Eastgate Hotel	Oxford 694	0865-248244
Mitre Hotel	Oxford 335	0865-244563

The dialling codes themselves contain smaller groups of figures for the larger cities, and in the late 1980s were:

01	London
021	Birmingham
031	Edinburgh
041	Glasgow
051	Liverpool
061	Manchester
091	Newcastle upon Tyne (and district)

A typical London number thus consists of 01, plus a three-figure number representing the former named exchange, plus a four-figure number for the telephone's location, giving nine digits in all.

But the demand for telephone lines in London is so great, that in 1990 the present simple 01 code will be replaced by two separate codes.The so far unallocated dialling codes 071 and 081 (see above) will be used for Inner and Outer London respectively.

Meanwhile, certain three-figure numbers are in use for dialling particular telephone services, such as 100 for the Operator, 151 to report a fault, 155 for the International Operator, 192 for Directory Enquiries, and 150 for British Telecom itself.

Easily the best known telephone number in Britain, however, is 999, used for the emergency services. And until fairly recently, another familiar number was the one for Scotland Yard: Whitehall 1212 (now 01-230 1212).

The final four digits of some London numbers are still memorable for being in exact thousands, repeated digits, or sequential numbers. Among some of the famous ones are:

The Home Office	3000
House of Lords	3000
The Daily Telegraph	5000
Harvey Nichols	5000
Daily Mail	6000
Daily Express	8000
Science Museum	8000
Post Office	8000
US Embassy	9000
Ministry of Defence	9000
Channel 4 Television	4444
British Telecom Head Office	8888
Harrods	1234
Army and Navy Stores	1234
Liberty's	1234
Central Office of Information	2345
Daily Mirror	0246
Board of Inland Revenue	6420

But Buckingham Palace, for example, is (at least to the public), simply 01-930 4832.

COINS AND CURRENCIES

Many coins of the world, past and present, have evolved number names denoting multiples of a particular currency. Here are some of the best known, grouped by value, together with their country of origin:

2 dobla (Spain), dobler (Majorca), doblone (Italy), dobra (Portugal), doppia (Italy), double (France), doubloon (Spain), duarius (Hungary)
3 dreier (Germany), drielander (Holland), triens (Ancient Rome), trime (USA), triobol (Ancient Greece), trojak (Poland)
4 chetvertak (USSR), cuartilla (Mexico), cuartillo (Spain), cuarto (Spain), cuatro (Bolivia), farthing (England), ferding (Sweden), fyrk (Sweden), kwartnik (Poland), örtli (Switzerland), quadrans (Ancient Rome), quart (France), quarter (USA), quarto (Italy), vierer (Switzerland)
5 fünfer (Switzerland), quinarius (Ancient Rome), quincussis (Ancient Rome)
6 sechser (Germany), seiseno (Spain), sixain (France)
10 décime (France), decussis (Ancient Rome), denarius (Ancient Rome), dime (USA), dinar (Algeria, Yugoslavia), dinero (Spain), dizain (France), zehner (Prussia)
100 cent (France, USA), centime (France), centimo (Venezuela), qindar (Albania), stotinka (Bulgaria)
1000 mil (Cyprus), milesimo (Chile), mill (USA), millime (Tunisia), milreis (Portugal, Brazil)
Several other coins have names denoting intermediate values.

ROADS AND ROUTES

All Britain's major **roads**, and most minor roads also, are officially numbered, with the best-known numbers being those of the motorways.

Road route numbers were introduced in 1920,

with the rapid advance of motoring and motorized transport generally. In that same year, Lord Montagu of Beaulieu produced a design for a London to Birmingham motorway that bore an uncanny resemblance to the plan ultimately adopted for the M1 itself, which opened in 1959. (It was not Britain's first motorway, however: that was the Preston bypass, now a section of the M6, which had opened the previous year.)

By the end of the 1980s, the main **motorway** routes in Britain had become familiar from their respective numbers:

M1 London–Yorkshire
M2 London–Medway Towns (Kent)
M3 London–Basingstoke
M4 London–South Wales
M5 Birmingham–Bristol–Exeter
M6 Birmingham–Carlisle
M8 Edinburgh–Glasgow–Greenock
M9 Edinburgh–Stirling
M11 London–Cambridge
M18 Rotherham–Goole
M20 London–Folkestone
M25 London Orbital Route
M40 London–Oxford
M62 Lancashire–North Humberside

The motorway network in England and Wales is based on five of these: M1, M4, M5, M6 and M62.

In many cases, the motorway number is the same as that of the trunk road that runs (or ran) on the same route. The A1, for example, formerly known as the Great North Road, still extends from London to Yorkshire. On the other hand, the A3 does not run through Basingstoke, but follows a route London–Guildford–Portsmouth.

Other countries have similar motorway systems. France, for example, has recently opened several 'A' motorways (*autoroutes*) to supersede the former 'N' roads (*routes nationales*), so that the A6 runs south from Paris to Lyon, and the A7 from Lyon on to Marseille. In the USA, the Interstate Highway System, begun in the 1950s, carries similar numbered routes, as do other main highways. One well-known such road is Route 95, which runs down the entire eastern seaboard, from the border with Canada south, through New York, to Miami, Florida.

AUTONUMEROLOGY

Roads means cars, and cars means **registration numbers**.

In Britain, car registration numbers were introduced in 1903, and the Road Traffic Act that year stated that all motor vehicles were required by law to be registered with a distinctive number and that they were to display that number on number plates.

The first registration number was not A1, as is sometimes supposed, but is believed to be CA1. This number was reserved in September 1903 for the MP who guided the Road Traffic Act through Parliament. A1 was allocated in December 1903 to Earl Russell, a former Under-Secretary for Air, who is said to have sat up all night to be first in the queue for this particular number.

Ever since car numbers were first allocated, many car owners have become keen, almost manic collectors of 'cherished numbers', that is, of letter and number combinations that reflect their individuality. The lower the number, the greater its prestige (and actual) value. And if a low or significant number can be combined with a meaningful letter or letters, so much the better! Today, autonumerologists, as they style themselves, are prepared to go to endless trouble and to pay large sums of money to acquire the number they desire, and there are specialist firms trading in car numbers of this type for such enthusiasts.

Advertisements offering distinctive car numbers appear regularly in the press, especially in such newspapers as the *Sunday Times*, and desirable numbers can fetch literally thousands of pounds.

Here are some examples of 'cherished numbers', together with their asking price, as advertised in a recent *Sunday Times*:

6 ACA	£2995
AT 10	£5000
BBC 1	£27,990
C 5	£45,000
EV 1	£19,800
1 FS	£22,300
HHR 36K	£80
LAM 1	£30,000
LOT 115	£17,000
MAG 1	£30,000
MY 1	£25,000
PS 2	£25,000
SS 8	£14,000
VEE 1	£14,000
YZ 1	£17,500

FLAG NUMBERS

Many national flags display a particular number of objects or symbols to represent a particular historic or geographical association. One of the best examples is the American flag, the 'Stars and Stripes'. The current flag has 13 horizontal stripes (seven red alternating with six white), with a blue canton charged with 50 white, five-pointed stars arranged in nine rows (six stars alternating with five). The stripes represent the 13 original founder-states of the North American Union, while the stars stand for the present total of member states. The present American flag is the 27th version, dating from 1960. The first American flag with 13 stripes was created in 1775, but its canton contained the British Union Flag. The Malaysian flag is very similar to the American flag, and was directly inspired by it. However, it has a star and crescent in its blue canton, and 14 horizontal stripes, not 13, representing the 13 member states of the Malaysian Federation and the territory of its capital, Kuala Lumpur.

An interesting aspect of autonumerology is the recognized endeavour to utilize numbers as letters of the alphabet, in order to form a meaningful word, especially the owner's name.

Thus, a lady named Mrs Parish displayed her car number, PAR 15H, in such a way that the '1' suggested an 'I' and the '5' an 'S', spacing out the number overall to read P A R 1 5 H. (The deliberate incorrect display of a number like this is a legal offence, and the practice is one that the police are constantly on the alert to detect.)

The traditional misinterpretations of figures as letters usually involves the following readings:

0 as O
1 as I or L
5 as S
6 as G
7 as T
8 as B
13 as B

In addition, numbers can be understood as words:

1 as one, won
2 as to, too
4 as for, fore
8 as ate

The number 20 FUS can thus be (illegally) spaced to read 'two of us', and one lucky owner proudly displayed his CLA 55Y plate to show how C L A 5 5 Y he and his motor were.

Show business personalities are particularly prone to 'meaningful' car numbers of this kind, and a typical sampling will be found in the following selection of 'cherished numbers', as recorded in the pages of two books devoted to the subject: Noel Woodall, *Car Numbers: 1904–1974* (1974) and Tony Hill, *The Concise Guide to Car Numbers* (1983). It should be borne in mind, however, that many such numbers change hands several times, as one enthusiast buys from or sells to another, so that the current location or ownership of a given number may well not be as stated here:

A 1 owned by the Dunlop Tyre Company, Birmingham

A 7 the first royal car purchased by Edward VII

AA 10 formerly owned by the comedian Arthur Askey

AAA 1 formerly owned by the Marquis of Exeter, President of the Amateur Athletic Association

AD 1897 displayed on an 1897 Daimler

AGA 1 owned by Aga Cookers and Boilers Ltd

ALF 1 owned by Alf Bickerstaff, then by Alf Page

1 ANN presented to Princess Anne as a wedding present

B 111 owned by Bill Turner, a jeweller and motor-racing enthusiast

BET 1 owned by Ron Pickering, a turf accountant

BET 7 owned by Victor Lownes, of the Playboy Club

1 BMW displayed on a vintage BMW owned by the chairman of BMW Concessionaires

BOB 1 owned by Bob Charlesworth, who also owned 1 BOB

BRA 36 owned by a bra manufacturer, who claims the number indicates the ideal measurement

BS 6000 owned by Barney Slattery, an antique dealer; the number is his home telephone number

BUY 1 owned by a Guildford garage and motor dealer, who also owns BUY 3, BUY 12, BUY 1B and OWN 1

CCC 1 owned by Caernarvon County Council, and displayed on its original vehicle, a road roller

1 COD owned by Roy Fidler, a leading rally driver, nicknamed 'King Cod'

THE TEN-CODE

The ten-code (10-code) as used by CB-ers is itemized as follows in Chas Moore's *CB Language* (1981):

10-1	Poor reception
10-2	Good reception
10-3	Stop transmitting
10-4	OK, message received
10-5	Relay message
10-6	Busy, stand by
10-7	Leaving air
10-8	In service, subject to call
10-9	Repeat message
10-11	Talking too rapidly
10-12	Visitors present
10-13	Advise weather and road conditions
10-16	Make pick-up at...
10-17	Urgent business
10-18	Anything for us?
10-19	Nothing for you, return to base
10-20	My location is...
10-21	Call by telephone
10-22	Report in person to...
10-23	Stand by
10-24	Completed last assignment
10-25	Can you contact...?
10-26	Disregard last message
10-27	I am moving to channel...
10-28	Identify your station
10-29	Time is up for contact
10-30	Does not conform to FCC rules
10-32	I will give you a radio check
10-33	Emergency traffic at this station
10-34	Trouble at this station, need help
10-35	Confidential information
10-36	Correct time is...
10-37	Pick-up truck needed at...
10-38	Ambulance needed at...
10-39	Your message delivered
10-41	Please tune to channel...
10-42	Traffic accident at...
10-43	Traffic jam at...
10-44	I have a message for you
10-45	All units within range please report
10-46	Help motorist
10-50	Break channel
10-60	What is next message number?
10-62	Unable to understand, use landline
10-63	Network directed to
10-64	Network clear
10-65	Awaiting your next message/assignment
10-67	All units comply
10-69	Message received
10-70	Fire at...
10-71	Proceed with transmission in sequence
10-73	Speed trap at...
10-74	Negative
10-75	You are causing interference
10-77	Negative contact
10-81	Reserve hotel room for...
10-82	Reserve room for...
10-84	My telephone number is...
10-85	My address is...
10-89	Radio repairman needed at...
10-91	Talk closer to the mike
10-92	Your transmitter is out of adjustment
10-93	Check my frequency on this channel
10-99	Mission completed, all units secure
10-100	Stop at a lavatory
10-200	Police needed at...
10-2000	Dope pusher

COM 1C owned by the comedian Jimmy Tarbuck; the number is one of the best known 'cherished numbers' in Britain

CUE 1 formerly owned by Joe Davis, billiards and snooker champion

1 CUE owned by Jimmy White, professional snooker player

55 DAF displayed on a DAF 55 car

DEB 1 given as a present by her husband to Mrs Debbie Davies

DOC 99 owned by a Dr Iskander

DS 333 owned by Derek Segall, an ex-rally driver, whose telephone number is 3303 and whose lucky number is 3

1 EGG owned by the director of an egg-packing plant

EYE 1 owned by a contact lens manufacturer

1 EYE owned by Reg Haynes, a wealthy businessman who lost an eye as a tank driver in the Second World War

FAB 1 given as a birthday present by William Aston to his 'fabulous' wife

FAL 1C owned by John Entwistle, guitarist with the pop group, The Who, with the guitar regarded as a phallic symbol

FLY 1 owned by Lady Brabazon of Tara, wife of Lord Brabazon, the motoring and aviation pioneer

20 FUS formerly owned by singer and songwriter Tony Hatch and his girlfriend, and used as the basis of his song 'Two of us'

G 0 the official car of the Lord Provost, Glasgow (the second half of the number is the figure 0, not the letter O); the Lord Provost of Edinburgh has the car S 0, with 'S' the letter always used on Scottish numbers

GEE 29 owned by Brian Gee, who lives at 29 Pine Hill and who was born in 1929

GOT 1 formerly owned by A M Attlee, brother of the Prime Minister Clement Attlee

H 202 owned by the director of a company manufacturing hydrogen peroxide, the formula for which is H_2O_2

HOF 1 owned by Sir Hugh Fraser, chairman of the House of Fraser

HOP 1N displayed on a demonstration car (an Opel) used by a garage in Staffordshire; *The Times* of 22 November 1988 reported seeing the number on a Mercedes convertible in the City of London

HOW 1 owned by Peter How, chairman of the How Group, a Birmingham engineering company; he originally displayed the number to read HOW! before motorway police put a stop to this practice

HRH 1 presented to The Queen as a wedding present by the Royal Air Force

3 HRH owned by His Royal Highness Prince Ghazi Faisal III

JCB 1 owned by the company that makes JCB excavators, founded by J C Bamford; the company also owns many other JCB numbers, including JCB 2 to JCB 10

JET 1 displayed on the first turbine-powered Rover car, now in the Science Museum, London

JM 1 owned by John Moores, the founder of Littlewoods Pools; he also owns JM 2

K 7 owned by the Duke of Kent, who obtained it in 1957 when he was 7th in succession to the throne

18 K owned by a jeweller, who regards it as '18 carat'

KGB 1 owned by Trevor Sears, a London lawyer, who encounters constant disbelief from anyone who sees it

KGC 12D Formerly owned by Mrs Gertrude Shilling, the fashion designer and sporter of outrageous hats at Ascot; the first half of the number is of no significance, but the second represented her (pre-decimal) surname

LOT 2 owned by a chartered surveyor

LOT 4 owned by a Mr Bidwell, and displayed on his Lotus

10 MAN owned by the public relations officer of the Isle of Man Tourist Board, who spaced it to read 1 0 MAN

MAR 10 owned by a man whose first name was Mario

MMM 1 owned by Lord Montagu of Beaulieu, founder in 1952 of the Montagu Motor Museum

100 MPH owned by Anthony Crook, director of a firm of distributors of Bristol cars; his daughter, Carole Crook, owns AC 1; Anthony Crook also owns MPH 100

NAB 1 formerly owned by the Conservative MP, Sir Gerald Nabarro, who also owned NAB 2 to NAB 7 inclusive, as well as higher numbers

NHS 1 owned by Dr H W E White, a Scottish GP

NO 1 displayed on an Aston Martin, where it is designed to be understood as 'No. 1', not 'no one'!

NO 5 owned by Chanel Ltd (who of course are famous for their Chanel No. 5 perfume)

NOW 1 formerly owned by Earl Mountbatten of Burma

1 NUR owned by Michael Gerson; the number can be interpreted in various ways, such as 'inure' or 'in you are'; the car attracts attention in Germany, as German *nur* means 'only'; the same owner has NUR 1 and, more individually, GER 50N

OAO 1 originally displayed on the special Austin Healey owned by the racing driver Mike Hawthorn, where it was understood to be the 'one and only one'; later owned by the director of a paint manufacturing company, who displayed it symmetrically (the 1 under the OAO) on her Scimitar GTE

OK 1 owned by Charles Yeates, chairman and managing director of a large bus and coach manufacturing company

OMO 1 owned by the chairman of Procter & Gamble Ltd, the firm that manufactured OMO washing powder

ONE 1 the official car of the Vice Chancellor of Manchester University

ONP 1L quoted by *The Times* of 21 December 1988 as seen in the City of London on a Porsche with a woman driver

OO 1 owned by a Scottish ophthalmic optician

OXY 1 owned by the former chairman of the British Oxygen Company

PK 1 owned by the actress, Pat Kirkwood

POP 1 owned by the popular songwriter, Mitch Murray

1 PRO owned by the professional snooker player, Ray Reardon

PUB 1 owned by a Sussex licensed victualler

PYM 1 owned by a Kent family named Pym, who also own PYM 2 and PYM 3: PYM 4 was owned by the MP Francis Pym

RA 444 owned by Ruth Atkinson, and displayed on a Morgan 4/4

RAD 10 owned by the well-known actor and singer Tommy Steele, and previously by the radio personality, Bob Danvers-Walker, who later acquired the number TV 1

13 RAF owned by the Royal Air Force No. 13 Squadron at RAF Wyton

1 REV owned by a Surrey garage proprietor (not a clergyman)

RO 1 owned by Leslie Falkson, chairman of Regal Optical Industries and organizer of the famous annual London to Brighton Run for vintage cars; he regards it as the most regal number plate in the world—except that of The Queen, which bears no number at all!

RR 1 displayed on a Rolls-Royce owned by a firm of Rolls-Royce distributors, who regard it as the most valuable car registration number in the world

RU 18 the classic 'fun number' often seen in pubs (as a reminder of the legal age for drinking alcohol); the actual number is displayed on an E-Type Jaguar owned by a man who lives in Leeds

7 SEA the number was owned by Jimmy Chipperfield, the circus master; it was originally purchased by the Seven Seas Company, which imported exotic wildlife from countries around the world in order to stock Britain's wildlife parks

SKY 1 owned by Derek Douglas, of the Manchester Airport Hotel

777 SM owned by Stirling Moss, the famous racing motorist

T 8 formerly owned by the comedian Harry Tate, and appearing on the stage with him in his show

T 42 owned by a Dorset firm of tea merchants, who thus offer 'tea for two', as in the old song

TAX 1 owned by a firm of London minicabs, but changed hands several times since; in 1980 its owners took it (on a Land Rover) to Switzerland, where they were imprisoned overnight for allegedly parading a joke registration number

THE 7 displayed on an Austin 7

TT 1 owned for several years by the comedian Tommy Trinder

TV 1 owned by the radio and television personality, Bob Danvers-Walker (who earlier owned RAD 10)

V 40 owned by Vincent Forte, of the catering firm Trusthouse Forte

VET 1 owned by a Rotherham, Yorkshire, veterinary surgeon

VS 1234 owned by the famous danceband leader, Victor Sylvester; the number of course represents his 'lead-in' to the band

1300 VW displayed on a 1300 Volkswagen

WOE 1 owned by a funeral director from Birmingham

WON 140 won by a businessman in a raffle with winning ticket number 140

CB NUMBER TERMS

In addition to the ten-code, CB-ers use the following numerical code terms:

Double 18	two juggernauts side by side
Double 88s	love and kisses
Double 7	no, no contact
807	beer
18-wheeler	juggernaut
8s	best wishes
88s	love and kisses
5 and 9	clear strong radio signal
5-finger discount	stolen goods
5-2	uncertain, maybe (i.e. half 10-4)
40-footer	juggernaut
44s	love and kisses
40 roger	message received
4	(abbreviation of 10-4)
4-10	(emphatic 10-4)
Have a 36-24-36 tonight	have a good evening
Meeting 20	meeting place
On the 70	driving at 70 mph
Pair of 7s	no answer, negative contact
17-wheeler	juggernaut with a puncture
73	best wishes
10-10 till we do it again	goodbye

(from Chas Moore, *CB Language*, 1981)

WW 31 owned by a Mr W Woodhouse of Blackpool, who was born on the 3rd day of the 3rd month as a 3rd child; his daughter was born on the 1st day of the 1st month as a 1st child; his wife's birthday is on the 13th of the month; his family life is thus full of 3s and 1s
YOB 1 owned by the 'bovver boy' pop group, Slade

It will be seen that official cars frequently bear significant number plates, and this applies to most ambassadorial and mayoral vehicles. Here is a selection of **diplomatic numbers**, as used in Britain:

1 ARG Argentina
AUS 1 Australia
1 BE Belgium
BOT 1 Botswana
BG 1 Bulgaria
CAN 1 Canada
DOM 300 Dominican Republic
FIJ 1 Fiji (nice one!)
1 GER German Federal Republic
GRE 1 S Greece
HON 2 Honduras
HKJ 111 Jordan (Hashemite Kingdom of Jordan)
1 LES Lesotho
1 M Malaysia
NZ 1 New Zealand
NIC 1 Nicaragua
OMA 1N Oman
1 TON Tonga
USA 1 USA
1 SU USSR
ZAI 1 Zaire
ZAM 1 Zambia

AMERICAN AUTONUMEROLOGY

'Cherished numbers' are a much simpler pursuit in the **USA**, where the 51 states, each with its own rules, allows 51 identical number plates. Nor are there the legal problems encountered in Britain when acquiring or transferring a particular number, since a person simply applies to the state issuing office for the number (combination of letters and/or numbers) of his choice. So long as the number does not exceed 6 characters, it can read almost anything (within reason). Such personalized plates, unlike British ones, double as licences, and when the licence expires, either a new validity sticker is placed on it, or the plate itself is discarded. Discarded plates are sold by auction to enthusiasts, who value them according to their rarity. Since a number plate can contain letters only (for example, the owner's initial and name), there may well be no actual figures on the plate.

American personalized plates—called 'vanity plates'—are therefore different in concept from British 'cherished numbers'. Moreover, official cars, for reasons of security, carry standard issue plates, not special distinctive ones.

Even so, American car number plates were in use before British ones, and they were issued in New York State in 1901. However, car numbers were *first* issued in France, according to a Paris Police Ordinance of 14 August 1893, which stated: 'Each motor vehicle shall bear [. . .] the name and address of its owner, also the distinctive number used in the application for authorization. This plate shall be placed at the left-hand side of the vehicle—it shall never be hidden.'

SPORTING NUMBERS

Almost all games and sports, from the simplest game of cards to the most complicated ball game, involve the use of numbers, whether for the players, the rules, or the scores. A cricketer in an eleven, for example, can score single runs or hit a four or a six. A professional soccer player

GOLF BY NUMBERS

Golf clubs are known by both name and number. The names have come down from the earliest days of the game, but the numbers are an American invention, dating only from the 1920s. There are five wood clubs and ten irons, as follows:

	number	*name*
Woods	1	driver
	2	brassie
	3	spoon
	4	baffy
	5	(used instead of number 3 or 4 iron)
Irons	1	driving club, or cleek
	2	midiron
	3	mid-mashie
	4	mashie iron
	5	mashie
	6	spade mashie
	7	mashie niblick
	8	pitching niblick
	9	niblick
	10	wedge

can represent a team in any of the four divisions (I, II, III, IV) of the Football Association. A yachtsman can compete on (or in) a 505, a 470, a 420, or an 8-metre or 12-metre yacht. And a golf player can choose from a wide range of clubs, numbered 1 to 10.

Many sporting terms are quoted in the numerical entries for Chapter 8, but here are some of the most important score points and values in a dozen of the best-known sports and games.

1 *Association Football* scoring is by single goals.

2 *American Football* a touchdown scores 6, plus 1 for a conversion after it; a goal kick scores 3, and a 'safety' 2.

3 *Baseball* scoring is by single runs (attained when a player has successively touched 1st, 2nd and 3rd base and has reached home base).

4 *Billiards* 2 points for pocketing one's opponent's ball by hitting it with one's own, or for pocketing one's own ball off one's opponent's, or for a cannon; 3 points for pocketing the red ball by hitting it with one's own ball, or for pocketing one's own ball off red (an 'in-off').

5 *Cricket* scoring is by single runs, which can be run consecutively off a single ball, however; 4 for hitting the ball along the ground over the boundary; 6 for hitting the ball over the boundary without its touching the ground.

6 *Darts* each sector of the board has a different value, with a hit in the outer ring scoring double the sector score, and a hit in the inner ring (halfway between the outer ring and the bull) scoring treble that value; the bull itself is divided into two small concentric circles, with a hit in the outer scoring 25 and one in the inner 50; overall, scoring is by subtraction from the target total, which is 301 for a singles game, and 501 for a double.

7 *Dominoes* there are different games with the pieces, whose standard maximum number is 28, but scoring will depend on the values shown in the pips (not in figures) on each piece; each piece is divided into two halves, and in a complete set the 28 pieces represent all combinations of two numbers from the maximum of 6 to the minimum of 0 (blank), viz. 6/6, 6/5, 6/4, 6/3, 6/2, 6/1, 6/0, 5/5, 5/4, 5/3, 5/2, 5/1, 5/0, 4/4, 4/3, 4/2, 4/1, 4/0, 3/3, 3/2, 3/1, 3/0, 2/2, 2/1, 2/0, 1/1, 1/0, 0/0.

QUARTERS, NOT THIRDS

English shoe sizes increase in increments of one-third of an inch, and plans to switch to a more logical progression of quarter-inch increments have not yet been realized.

But 300 years ago, shoe sizes were measured in quarter-inch stages, as the following contemporary extract shows:

The Size of a shooe is the measure of its length which is in Children divided into 13 parts; and in Men and Women into 15 parts; the first of these being five inches long before it be taken for a size; what the shooe exceeds that length, every fourth part of an inch is taken for a size 1, 2, 3, and so forwards to 12 which is called the Boys or Girls Thirteens, or the short thirteens, and contains in length 8 inches and a quarter, from which measurement of 8 inches and a quarter the Size of Men and Women, called the long size or Man's size begins at 1, 2, 3, etc. to the number 15, each size being the fourth part of an inch as aforesaid; so that a Shooe of the long fifteens is in length 12 inches just.

(Randle Holme, *The Academy of Armory and Blazon*, 1688)

8 *Hockey* scoring is by single goals.

9 *Ice Hockey* scoring is by single goals.

10 *Rugby Football* 4 for a try; 6 for a converted try; 3 for a dropped goal, a goal from a mark, or a penalty goal.

11 *Snooker* the red ball scores 1, with the other colours as follows: yellow 2, green 3, brown 4, blue 5, pink 6, black 7; there are also penalties of different values, such as 7 for striking red when 'on' a colour, or for playing with the wrong cue ball, or 4 (or the value of the ball the player was 'on', if higher) for a miss.

12 *Tennis* the three main score points in a game are 15, 30 and 40, with a further 'advantage' point if both players are at 40–40 (deuce); the first player to win 6 games wins the set, but if the game score stands at 5–5, the next game is similarly an 'advantage' and play continues until one player has a lead of 2 games to win the set.

SPECIALIZED NUMBER SYSTEMS

However routine our everyday lives, at some point we are likely to encounter one or more kinds of specialized numbering systems. Here are just three of them, found respectively in the public library, the world of weather, and the City or financial pages of a daily newspaper.

WHAT'S YOUR IQ?

A number that interests many people is that of their IQ, or intelligence quotient. As we know them today, IQ tests were devised at the turn of the 20th century by the French psychologist Alfred Binet for measuring the mental ability of children. According to the scale he devised, a child's IQ was expressed by his mental age divided by his chronological (actual) age multiplied by 100. An average child thus scored 100. The IQ test can also be used on adults, but of course the concept of mental age breaks down, except for severely subnormal adults with, say, a mental age of 10. The child who to date has achieved the highest IQ score was an American ten-year-old, Marilyn de Vos. Her tests in 1957 produced a reading of 228, which is the ceiling score for a 23-year-old.

THE DEWEY DECIMAL SYSTEM

Many public and educational **libraries** in Europe and America use the Dewey Decimal System to classify their books, and the numbers of the different sections are normally marked out conspicuously for library users — whom they are designed to aid just as much as the librarians themselves.

The scheme is named after Melvil Dewey, an American librarian who devised his system in 1873 for use in the library at Amherst College, from where he was graduated the following year.

His system divides all knowledge into 10 groups, with 100 numbers allocated to each group (hence 'decimal'). Each group is quite distinct: religion, for example, is in the 200s, the arts in the 700s, and history and geography in the 900s. Within each group, the main sub-sections are divided by 10, so that books on the Christian religion (specifically) are numbered in the 260s, for example, and works dealing with geographical place-names are in the 910s. Books on numbers, like this one, are placed in the 510s. To each hundred-number, precise subdivisions can be indicated by using another figure after a decimal point. For example, the arts are in the 700s, the history of art is 709 and the history of art in the US is 709.73. The '73' is not random here, but is the specific hundred-number used for the history of the US (973), so that it acts as a mnemonic. Similar identity of numbers occurs elsewhere in the system, so that all books in a particular language, for instance, will have the same decimal classification.

Today, many books have their specific Dewey Decimal number printed on the reverse of the title page, as this one does.

Here are the Dewey numbers of some other books published by Guinness:

The Guinness Book of Records 031'.02
The Guinness Book of Names 929.9'7
The Guinness Book of Words 422
The Guinness Book of Almost Everything You Didn't Need to Know About the Movies 791.43'09

The sub-subdivision expressed for this last title indicates its highly specialized nature; compare the title above it, which is much more general so has simply the hundred-number, without any further need for decimal division.

WHATEVER THE WEATHER...

We are all affected by the **weather**, and most of us see or hear a weather forecast daily. In discovering whether it is to be hot or cold, wet or dry, windy or calm, or simple good weather or bad, we are presented with figures on at least three different measurement scales.

The first, or course, is the **temperature**, with the reading given in degrees Celsius or Fahrenheit.

The second is the **barometric pressure**, given in millibars. Broadly speaking we need to know that a reading of 1000 mb or more usually means fairly good or settled weather, while a figure

THE NUMBERS GAME

The numbers game, also known as 'numbers pool', 'numbers racket' or simply 'numbers', is a form of lottery popular in the US, where it is, however, illegal.

It is played mainly by blacks and low income groups, who bet around 10 cents on a three-figure number (from 001 to 999) of their own choice. The winning number is taken from a source that is outside the control of the players, such as stock market figures or betting pay-offs published in the press. The odds, obviously, are 998 to 1. Traditionally, the person who has selected the winning number is paid 540 to 1, with the 'runner', who takes his bet, getting 60 units and the gross profit of the promoters being nearly 40 per cent of the total bet. Thus, a 10-cent wager can land the lucky winner $54.

Promoters of the numbers game are known to pay politicians and the police large sums as protection money against arrest, and although several states run legal lottery systems, the numbers game is still as popular as ever.

under a thousand means bad or unsettled (stormy) weather.

The third is the amount of **rainfall**, measured in millimetres or inches. The precise amount of rain that will fall cannot be forecast, but rainfall readings are usually incorporated into a weather report. A fall of more than 50mm (1.97in) of rain over a 24-hour period is usually regarded as considerable.

A fourth important scale expresses the force of the **wind**. Wind force can be given either in miles per hour or according to the Beaufort Scale. The scale was devised by Admiral Sir Francis Beaufort in 1805 and now runs from 0 to 17 on an extended reading, although values of 0 to 12 are sufficient for most countries. The actual equivalents are as follows:

Scale number	*Description*	*Wind speed* (mph)	(knots)
0	Calm	1	<1
1	Light air	1–3	1–3
2	Slight breeze	4–7	4–6
3	Gentle breeze	8–12	7–10
4	Moderate breeze	13–18	11–16
5	Fresh breeze	19–24	17–21
6	Strong breeze	25–31	22–27
7	High wind	32–38	28–33
8	Gale	39–46	34–40
9	Strong gale	47–54	41–47
10	Whole gale	55–63	48–55
11	Storm	64–72	56–65
12	Hurricane	73–82	over 65

At the same time, we need to know that *gusts* of wind can occur at speeds of over 100mph. (Beaufort Scale 14 or higher), as they did in the violent storms of mid-October 1987 in southern England.

Our meteorological awareness thus needs to juggle and compare a whole range of different figures and values, to provide us with the overall weather picture that we want.

Here is the weather picture for London as recorded on 30 January 1989:

Maximum temperature 10°C (50°F)
Minimum temperature 4°C (39°F)
Barometric pressure at 6 p.m. 1042.9mb, rising
Rainfall nil
Hours of sunshine 1.6.

THE RICHTER SCALE

It is popularly supposed that the Richter Scale, on which earthquakes are measured, is a 'scale of 10', something like the Beaufort Scale for winds. In fact this is not the case, and the Richter Scale (named after the US seismologist Charles Richter, who died in 1985) is a scale for expressing the magnitude of a seismic disturbance, with the reading derived from a logarithmic plot. The scale is only a 'scale of 10' in that the logarithmic base of 10 is used, and it has no fixed upper or lower limit. As a coincidental matter of record, however, no earthquake has yet been indisputedly assigned a reading above 8.9 on the Richter Scale (although some estimates put the cataclysmic explosion at Krakatoa, in 1883, at 9.9). The name of the scale is frequently used humorously to describe an emotional shock or surprise, such as, 'When I heard I had passed the exam it was at least 8 on the Richter Scale'.

ANSWERS to 'Name that Number' questions on p. 67

1 98.6°F or 36.9°C
2 192 (142 in London)
3 1970
4 11
5 7
6 M3
7 CA1
8 700
9 25–31 mph
10 40 per cent
11 34 in
12 39 cm
13 75 mph
14 041
15 28
16 4
17 36-23-36
18 90
19 480 (or 500)
20 78 years
21 6
22 400°F or 204°C
23 16
24 7 (or 13½)
25 5 ft 4 in (1.62 m)

8 NUMBERS IN LANGUAGE AND LITERATURE

NUMERICAL EXPRESSIONS AND QUOTATIONS

Numbers occur in a wide range of everyday expressions and familiar quotations, and this chapter is devoted to a selection of each.

In numerical order, examples are given of expressions, and quotations and sayings, with the latter including, where known, the name of the author and the title and date of the quoted work.

Obviously, the lower the number, the greater the range of expressions, especially as 'oneness' includes not just one and once but first, prime (and primary) and single, while 'twoness' covers not simply two and twice but second (and secondary) and double.

In many cases, number expressions will have a literal 'opposite number', or contrast, so that many primary objects have a paired secondary. In some cases, too, there may be a third member, which here would be a tertiary. Readers should follow up such contrasts at leisure. But they will mostly not be indicated, as to cross-refer to every 'opposite number' would simply take up too much space. In a few cases, however, a note to 'compare' or 'see also' is provided to denote a related expression.

Although the range of numerical expressions quoted is fairly comprehensive, the quotations and sayings are much more selective and are therefore restricted to a small range of examples. Since **ages** are important to everybody, however, quotes referring to ages usually have precedence, especially in the 'rarer' numbers.

> **LILLIPUTIANS**
>
> In *Gulliver's Travels* the Lilliputians are only six inches high, and thus 12 times smaller than sixfooter Gulliver. The area occupied by the Lilliputians is 150 times (12 X 12) smaller than ours, while the volume they take up is 1700 times (12 X 12 X 12) less.
>
> Jonathan Swift exploited the Lilliputians' tininess to create many amusing situations and misunderstandings, while using their smallness to satirize the pettiness of (real-size) human beings.

ONE

EXPRESSIONS

BASED ON **ONE** AND **ONCE**

A1 excellent (from the ranking used in Lloyd's Register of Shipping, where 'A' relates to the hull of the ship and '1' to her stores).

back to square one back to the beginning after making some progress (probably from a board game such as snakes and ladders, where a player can be sent back to the first square, where he started).

committee of one a person who appoints himself as the sole authority.

Formula One a racing car of a prescribed engine capacity, size, weight, etc., competing in a Grand Prix race; the formula (or specification) is frequently adjusted to keep rising speeds safe.

Jimmy the One an alternative name for **Number One** (4).

like one o'clock quickly, excellently (said to be from speed with which factory workers leave for 1.00 lunch break).

murder one a charge of **first-degree murder**.

No. 1, London the official address of Apsley House, Piccadilly, former home of the Duke of Wellington.

number one (1) the speaker, as the most important person; (2) urine (children's slang); (3) the best obtainable ('a No. 1 bestseller'), the most popular ('No. 1 in the charts'); (4) (Number One) the **first lieutenant** on a ship.

numbers 1s a sailor's best uniform.

numero uno the best or most important person or thing (Italian).

once and for all finally, for good.

once in a blue moon very rarely (because the moon is normally yellow or silver, not blue).

oncer (1) a (former) £1 note; (2) a person who goes (or used to go) to church once only on Sunday (compare **twicer**).

once removed (a cousin who is) the child of a person's **first cousin**.

one and only a unique person, especially a sweetheart.

one-armed bandit a gambling or 'fruit' machine, which has one arm or lever and which usually 'robs' its players.

one degree under feeling rather 'low' (slogan of Aspro ad in 1960s).

one-design yacht built to a standard design, so that it is identical to others.

one-eyed inferior (originally of a horse with one eye).

one fell swoop a single, concentrated go or effort (from Shakespeare, *Macbeth*, where 'fell' originally meant 'evil', as in modern 'felon').

one foot in the grave half-dead, very ill.

one for the book a deed or fact worth noting.

one for the road a final drink before leaving for home.

one-horse town a small town with few amenities (and originally one where the community had just one horse).

one-house bill in US, a piece of legislation passed either by the House of Representatives or the Senate for public-relations purposes; it never becomes law, and is effectively killed as a bill.

one-inch map a map with a scale of 1 in = 1 mile, formerly used for Ordnance Survey large-scale maps.

one-in-ten hill an incline with a gradient of one foot vertically for every 10 horizontally.

one in the eye a (figurative) blow in the eye, a setback or letdown.

one-legged lacking something essential (a 'one-legged rule' has no basis or logic).

one-liner a short joke or witty remark (originally a headline in a single line).

one-man band a group of instruments played simultaneously by one person, a group of activities run by one person without any help.

one-man dog a dog loyal only to his master, a person loyal only to his employer or superior.

one-night stand (1) a single evening performance, especially in a town where there is an audience for only one night; (2) a casual sexual relationship lasting one night only.

one off a single occurrence or example of something.

one-one (1:1) a Cambridge honours degree in the 1st section of the 1st class (compare **two-one**).

one on one person to person, in direct confrontation.

one-parent family a family lacking either father or mother.

one-piece a garment consisting of a single piece (compare **two-piece**, **three-piece**).

one-pipper a second lieutenant in the army (who has only one shoulder pip).

oner a person or thing regarded as unique or outstanding (the word happens to suggest 'wonder' and 'won').

one-reeler a short film on a single reel, usually lasting only 10 minutes.

one-ringer a flying officer in the RAF (who has a single ring on his sleeve).

one-sided argument an argument that is not reciprocal.

one-step a ballroom quick-time dance with single walking steps backwards and forwards (compare **two-step**).

one thing at a time slowly but surely.

one-time pad a writing pad used once only for a cryptic (coded) message.

One Ton Cup an annual international yacht race, originally held for yachts in the one-ton class, but now restricted to yachts not exceeding 22 ft (6.7 m) in length.

one-to-one a matching relationship of two people or things.

one-track mind a mind with a single dominant thought or obsession (like a single railway track, which has traffic in one direction only).

one-two (1) punches with alternate fists in boxing; (2) a football pass where one player kicks the ball to another then runs ahead for it to be kicked back to him.

one up superior (as if having gained an advantage or won a point in a game) (compare **one down**); the pursuit of social superiority is known as oneupmanship.

one-way mirror glass that can be looked through from one side but which on the other side appears to be a normal mirror (compare **two-way mirror**).

one-way street a street with traffic in one direction only.

one-way ticket a single ticket, for a journey in one direction only.

to be one down in an inferior position, as if having lost a point in a game (compare **one up**).

to go one better to improve on someone else's performance (from poker, meaning 'raise the stakes by one').

V1 a German flying bomb used in the Second World War (short for *Vergeltungswaffe*, 'retaliation weapon') (compare **V2**).

BASED ON **FIRST**

City of Firsts Kokomo, Indiana (as the first to introduce the mechanical cornpicker, the push-button car radio, the commercially-built car, and canned tomato juice).

first aid emergency medical assistance.

1st Amendment (of US Constitution) freedom of religion, speech, the press, assembly, and petition; with the 2nd to 10th Amendments forms the Bill of Rights (1791).

first base in baseball, the base that must be touched first by a batter when attempting a run (compare **get to first base**).

First Cause God, as the (uncaused) creator of all things, apart from himself (compare **Prime Mover**).

First Church of Christ, Scientist the mother church of Christian Science, Boston, where founded by Mary Baker Eddy in 1879.

first class the highest status of service, ability, etc.

first course the main (savoury) dish of a meal, typically based on meat or fish.

first cousin the child of a person's uncle or aunt.

First Covenant God's promise to Noah that there would never be another flood to cover the earth (Bible: *Genesis* 9:9).

First Day the name of Sunday, as (formerly) used by Quakers.

first-day cover an envelope postmarked with the date of issue of a new stamp or stamps.

first-degree burn a mild burn, with the skin painful and red but not more seriously damaged.

first-degree murder murder by poison or premeditated design, punishable by death in many US states.

first edition the original issue of a book, often later regarded as the most valuable, especially by collectors.

First Empire the imperial rule in France under Napoleon Bonaparte (1804–14).

first estate the clergy of a country, especially the Lords Spiritual in England or the pre-revolutionary clergy in France.

first family the family of the US President, or of a state governor (compare **First Lady**).

First Fleet the fleet that arrived at Port Jackson, Australia, in 1788, from which year the country dates its foundation.

first floor the ground floor (in US), or the one above the ground floor (in UK).

First Folio the first published collection of Shakespeare's plays in 1623.

first-footer a person who first enters a house on New Year's Day, especially in a country (such as Scotland) where the New Year is celebrated.

First Four Ships the four ships of English settlers that arrived at Lyttelton, New Zealand, in 1850, so founding Canterbury.

first-fruits the first products of one's efforts, as if the 'fruit' of what one has 'sown'.

first gear the lowest gear in a motor vehicle or on a bicycle, used for starting or climbing a gradient.

first generation the earliest, most primitive model of a machine or equipment; of computers, regarded as those built between 1948 and 1956.

first-hand direct, new, original, as a 'first-hand report' (compare **second-hand**).

First International the association of socialists and labour leaders founded in London in 1864.

First Lady the wife of the US President, or of a state governor (compare **first family**).

first law of motion the law of inertia – if a body is at rest or moving at a constant speed in a straight line, it will remain so until acted on by a force (Newton, 1687).

first law of thermodynamics the law of conservation of energy – the change in the internal energy of a system is equal to the sum of the heat added to the system and the work done on it (Mayer, 1842).

first lieutenant the executive officer of a warship (not necessarily a lieutenant in rank), especially a smaller one in the Royal Navy, also an officer of commissioned rank in the US Army, Air Force, or Marine Corps, senior to a **second lieutenant**.

First Lord of the Admiralty the former parliamentary head of the Royal Navy; in 1964 the Admiralty was reorganized as the Ministry of Defence (Navy).

first mate (or **first officer**) the assistant to the captain of a ship (compare **second mate**).

first mortgage a mortgage that has priority over all others.

first name a person's forename or Christian name, belonging to him or her individually (as distinct from a family name).

first offender a person who has been convicted of a criminal offence for the first time.

first past the post a candidate in an election who wins by a simple majority, not an absolute majority (term borrowed from horseracing, where the first horse past the post is the winner).

first person in grammar, the term referring to the speaker or speakers (I, we).

first position (1) the lowest possible position of a player's left hand on the fingerboard of a stringed instrument; (2) in ballet, the position with heels together, and toes turned out so that the feet are wide-splayed, while the arms are curved inwards at chest level.

first quarter the appearance of the Moon as a half-moon, approximately one week after the new moon.

first rate excellent (term originated from rating of a ship by the number of guns she carried and by their weight; there were six rates, with the first rate the highest).

first reading the introduction of a bill in Parliament.

first refusal the chance to buy or do something before another person.

First Reich the medieval Holy Roman Empire (from the German word for 'kingdom').

First Republic the republic in France from the abolition of the monarchy (1792) to the coming to power of Napoleon (1804).

first school a school for children aged 5 to 8 or 9 (compare **primary school**).

first sergeant a senior non-commissioned officer in the US Army.

First State Delaware (as the first to ratify the US Constitution, in 1787).

first strike the initial, usually pre-emptive attack in a nuclear war (compare **second strike**).

first string the top players of a team in an individual sport, such as squash.

A LOVER'S EXTRAVAGANT COUNT-UP

The 17th-century parson and poet Robert Herrick wrote several simple but passionate love poems. Some of the best known are those dedicated 'To Anthea' in his collection *Hesperides*. In 'Ah, My Anthea!' he multiplies his kisses by leaps and bounds:

> Give me a kiss, add to that kiss a score;
> Then to that twenty, add a hundred more:
> A thousand to that hundred: so kiss on,
> To make that thousand up a million.
> Treble that million, and when that is done,
> Let's kiss afresh, as when we first begun.

These lines closely echo those of the Roman poet Catullus, written in the 1st century BC:

> *Da mi basi mille, deinde centum,*
> *Dein mille altera, dein secunda centum,*
> *Deinde usque altera mille, deinde centum.*

('Give me a thousand kisses, then a hundred, then another thousand, then a second hundred, then yet another thousand, then a hundred.')

First Triumvirate the coalition between Pompey, Caesar and Crassus in 60 BC.

first watch the 8.00 p.m. (2000 hours) to midnight watch on a naval ship.

first water the highest quality of a diamond ('water' indicating its degree of brilliance).

First World War the global war of 1914–18, originally known as the Great War (compare **Second World War**).

Glorious First of June 1 June 1794, the date of the British naval victory over the French in a battle in the Atlantic west of Ushant (Ouessant), north-west France.

to get to first base to carry out the first stage of an enterprise or activity (from baseball: see **first base**).

BASED ON **PRIME** OR **PRIMARY**

primary accent the main stressed syllable in a multisyllabic word, such as the 3rd syllable in 'consti*tu*tional' (compare **secondary accent**).

primary cell a cell that produces an electric current by an irreversible chemical action, so that it cannot be recharged.

primary colour a colour from which all others can be derived, especially red, green or blue.

primary election in the US, the election of the local members of a party, or of the delegates who will nominate them.

'FORTY YEARS ON'

The famous Harrow School Song, 'Forty Years On', was written in 1872 by a Harrow schoolmaster, Edward Ernest Bowen, to mark the tercentenary of the school's foundation.

The title, which also forms the opening words, suggests Dumas' novel *Twenty Years After*, but the number 40 is purely arbitrary, meaning 'many'. Even so, the words could be said to apply to a middle-aged man in his mid-50s who was once a pupil at Harrow, and verse 4 of the song has the expanded wording: 'Twenty, and thirty, and forty years on'.

There is a specific reference to Harrow Football in verse 1, where the words describe the 'tramp of the twenty-two men'. Harrow Football has 11-a-side, as does Association Football, but differs from it in that the distinctive ball, like a medicine ball, can be both kicked and handled.

A sixth verse was added to the song in 1972 to mark the school's quatercentenary, with the first line running: 'Four hundred years in the story of Harrow'.

The song title was later adopted by Alan Bennett for his first play (1968), about a minor public school where the boys enact a satirical pageant of recent British history in honour of the retiring headmaster.

primary planet one of the 'main' planets, such as Mars or Earth, as distinct from a satellite or minor planet.

primary school one for children aged 8 to 11 (compare **first school**).

prime minister the head of state, the 'first minister'.

Prime Mover God, as the **First Cause** (i.e. the 'uncaused' creator of all things, apart from himself).

prime number a number that can be divided exactly only by itself and one.

BASED ON **SINGLE**

single (1) a score of one run in cricket or one point in some other game; (2) a gramophone record with one song (piece) on each side.

single bed a bed for one person.

single-breasted (of a coat or jacket) with fronts slightly overlapping and one row of buttons or fasteners.

single cream cream with a low fat content, that does not thicken when beaten.

single-decker a bus with one deck, or any object that could be multi-decked, such as a cake, but is not.

single entry in bookkeeping, a system by which transactions are entered on one account only.

single file a line of people, animals, etc. one behind the other.

single-handed done by one person, without the help of anyone else.

single life an unmarried life, lived on one's own.

single market the unified market of the European Economic Community that will operate from 1992.

single parent an adult who looks after a family alone.

single portion a serving of food for one person.

single-sex designed for one sex only, such as a school.

single-space typing done without a space between the lines.

single ticket a ticket for travel in one direction only.

singleton one card in a suit (in bridge) (compare **doubleton**).

single track a railway or road along which transport travels in one direction only.

QUOTATIONS AND SAYINGS

WITH **ONE** AND **ONCE**

Ein Reich, Ein Volk, Ein Führer (One realm, one people, one leader). (German Nazi slogan, Nuremberg rally, 1934)

E pluribus unum (One out of many). (motto on the face of the Great Seal of the US, adopted 1792 from Virgil)

Genius is one per cent inspiration and ninety-nine per cent perspiration. (Edison, *Life*, 1932)

In the kingdom of the blind, the one-eyed man is king. (H G Wells, *The Country of the Blind*, 1911)

My lucky number's one. (Lene Lovich, pop song 'Lucky Number', 1979)

Once bitten, twice shy. (proverb)

One a penny, two a penny, hot cross buns. (nursery rhyme)

One for all, all for one (*Tous pour un, un pour tous*). (motto of the Three Musketeers in Dumas' novel, 1844, but quoted much earlier by other writers, such as Shakespeare)

One good turn deserves another. (proverb, but quoted by Gaius Petronius in 1st century AD)

One if by land, and two if by sea. (the signals agreed on by Paul Revere to warn of the arrival of the British forces in 1775, quoted by Longfellow in *Paul Revere's Ride*, 1861)

One step forward, two steps back (*Shag vperëd, dva shaga nazad*). (title of book by Lenin, 1904, condemning party disunity: the 'one step forward' of the Bolsheviks was countered by the 'two steps back' of the Mensheviks)

One man, one vote. (electioneering slogan, originally in form 'One man shall have one vote', John Cartwright, *People's Barrier Against Undue Influence*, 1780)

One's impossible, two is dreary,
Three is company, safe and cheery
(song from Stephen Sondheim's musical *Company*, 1970)

One swallow does not make a summer. (proverb, originally quoted by Aristotle, 4th century BC)

That's one small step for a man, one giant leap for mankind. (Neil Armstrong, on being the first man to set foot on the Moon, 1969)

The Lord our God is one Lord. (Bible: *Deuteronomy* 6:4)

WITH **FIRST**

Eclipse first, the rest nowhere. (Dennis O'Kelly, 1796, when the great racehorse Eclipse was about to run his first race at Epsom)

First come, first served. (proverb)

First in war, first in peace, and first in the hearts of the citizens. (Henry Lee on the occasion of George Washington's death in 1799; 'citizens' was later changed to 'countrymen')

First things first. (proverb)

First things first, second things never. (Shirley Conran, *Superwoman*, 1975)

I had rather be the first man in a village than second at Rome. (quoted by Bacon, *The Advancement of Learning*, 1609)

The evening and the morning were the first day. (Bible: *Genesis* 1:5)

The first blow is half the battle. (Goldsmith, *She Stoops to Conquer*, 1773)

The first step is the one that is difficult (*Il n'y a que le premier pas qui coûte*). (Mme du Deffand, letter to d'Alembert, 1763, commenting on the legend that St Denis walked two leagues carrying his head in his hands)

TWO

EXPRESSIONS

BASED ON **TWO** AND **TWICE**

beast with two backs sexual intercourse (old idiom).

fall between two stools to succeed in neither of two actions (from old proverb, dating at least from 14th century).

Goody Two-Shoes nickname for a goody-goody (from an old fairy tale about a girl who had only one shoe before being given another, which she smugly showed everyone).

in two minds undecided, uncertain.

in two shakes of a lamb's tail very quickly.

in two twos very quickly (perhaps originally referring to the speed with which the two times table is learnt or said).

kill two birds with one stone to carry out two tasks or solve two problems with the effort needed for only one.

mark II a second, usually improved model of something.

not to give two hoots not to care at all ('hoot toot' was formerly an exclamation of displeasure or impatience).

no two ways about it unequivocal, unambiguous, definite.

number two (1) faeces (children's slang); (2) a second-in-command (compare **number one** (4)).

number two town in theatre jargon, a provincial town, with a smaller audience than in London or a major city.

put two and two together to come to a correct conclusion.

Shrewsbury Two Eric Tomlinson, Dennis Warren: builders jailed (1973) for violence on building sites in Shrewsbury.

stand on one's own two feet to be independent, without the help or support of anyone else.

thick as two short planks obtuse, dim-witted.

twice-born 'reborn', regenerate (for example as a 'born-again' Christian).

twice-laid made of old yarns that have been retwisted.

twicer (1) a printer who works as both compositor and pressman (disparaging nickname); (2) a person who goes (or used to go) to church twice on Sunday (compare **oncer**).

twice-told of a story: recounted twice, hackneyed.

two-and-a-half-ringer squadron-leader in RAF (who has two broad rings and one thin one on his sleeve).

two a penny plentiful, commonly found or occurring.

two-bit cheap, almost worthless ('two bits' is a small sum of money: in US, 'bit' = 25 cents, or quarter of a dollar).

two-by-four a length of timber with a cross-section 2×4 in (compare **four-by-two**).

two-car family a family running two cars.

two cents' worth a small amount.

two cheers qualified applause or approval (compare **three cheers**).

two cultures the sciences and the arts (from title of book by C P Snow, 1956).

two-dimensional relating to length and breadth, or length and height.

two-dog night a very cold night (from the number of sled dogs needed to keep an Arctic traveller warm at night).

two-earner a family where both husband and wife have full-time jobs.

two-edged remark an ambivalent or ambiguous remark.

two-faced insincere, hypocritical.

twofer (1) in US, a cheap cigar (originally selling at two a nickel); (2) a theatre ticket sold at half the normal price, that is, two for the normal price of one.

two fingers a 'V-sign', as a disrespectful or defiant gesture.

two-four time in music, a measure with two crotchets in a bar, a crotchet being a quarter-note.

two-hander a play for two actors.

two irons in the fire an available alternative if one choice fails (originally, a spare hot iron to use if one is not hot enough).

two left feet awkward, clumsy.

two-line whip in politics, an instruction to MPs to attend a debate and vote as told (with the instruction underlined twice) (compare **three-line whip**).

2LO the call sign of the BBC in the early years of broadcasting ('LO' for London).

two minutes' silence a silent tribute to the dead, as on Armistice Day or Remembrance Sunday.

two-name paper in US, a commercial paper signed equally by two persons.

Two Nations the rich and the poor (from the secondary title of Disraeli's novel *Sybil*, 1845).

two natures in theology, the divine and human natures that are united in Christ.

two-one (2:1) a superior Class 2 honours degree at Oxford or Cambridge (compare **one-one**).

two or three a few.

two-party system having two major political parties, as the former British Whigs and Tories, or Conservative and Labour, or the US Republicans and Democrats.

twopence (tuppence) an amount of little value, as 'not worth twopence').

twopenny-halfpenny of little value, as a 'twopenny-halfpenny job'.

Twopenny Tube a former nickname for the Central London Railway, on which the fare was 2d for some years after opening (1900) ('Tube' as ran underground through tunnels, like present London Underground).

two-piece a garment or costume in two pieces, such as jacket and trousers (slacks), or a woman's swimsuit (compare **one-piece**, **three-piece**).

two-pot screamer in Australia, a person easily influenced by alcohol.

twos and threes a children's chasing game for six or more players (who are in groups of two or three).

two-seater a vehicle with just two seats, one for the driver and the other for a passenger.

Two Sicilies the kingdom of southern Italy that existed from 1061 to 1860, comprising Sicily and Naples.

two-sided argument a controversial argument.

twosome a game for two.

two-step a dance in **two-four** or **four-four** time with a basic pattern of two steps ('step, close, step') (compare **one-step**).

two-stroke (two-cycle) a combustion engine with a piston making two strokes for each explosion.

two-time to deceive, **double-cross**, especially of a lover who carries on a relationship with someone else.

two-tone: (1) consisting of two colours, or giving two notes; (2) a type of black popular music that is a mixture of reggae and new wave.

Two-Ton Tessie the nickname of the amply-proportioned British music-hall singer, Tessie O'Shea (born 1917), or of any woman of similar build (compare **Two-Ton Tony**).

Two-Ton Tony the nickname of the generously-built American heavyweight boxer Tony Galento (1910–79), or of any man of similar adiposity (compare **Two-Ton Tessie**).

two-two (.22) a gun, or the ammunition for it, with a bore of .22in (5.6mm) (compare **twenty-two**).

two-up in Australia, a gambling game in which two coins are tossed or spun, with bets made on which side will land uppermost.

two-up, two-down a small house with two bedrooms upstairs and two reception rooms downstairs.

two-way mirror a mirror that is normal on one side but that can be seen through on the other, making it suitable for secret observation without being seen (compare **one-way mirror**).

two-way stretch an elastic garment that stretches in both directions, i.e. both length and width.

two-way switch a switch that operates in two places to control a single device, such as a light switch at the top and bottom of a staircase.

two-way traffic a road or track on which transport can run in both directions.

V2 the rocket-powered ballistic missile used against Britain by the Germans in the Second World War (initial of German *Vergeltungswaffe*, 'retaliation weapon') (compare **V1**).

wear two hats to hold two different positions of responsibility.

BASED ON **SECOND** AND **SECONDARY**

Deuxième Bureau the former French Intelligence Department.

Second Adam Jesus, the equivalent of Adam, the first man (compare **Second Eve**).

Second-and-a-half International the international workers' union of socialist parties set up in 1921 in Vienna by centrist socialists of several European countries and the USA, together with the Russian Mensheviks (so nicknamed as falling between the **Second International** and the **Third International**; also called the Vienna International).

secondary accent the second strongest syllable in a multisyllabic word, such as the first syllable in (*con*stitutional' (compare **primary accent**).

ELIOT'S FOUR QUARTETS

T S Eliot's long philosophical poem, *Four Quartets*, is in four parts, published as a whole in 1943. The title is intended to suggest a musical work, and obliquely refers to Beethoven's last four string quartets. The four parts are named after four different places—*Burnt Norton*, *East Coker*, *The Dry Salvages*, *Little Gidding*—and are respectively concerned with the four elements (air, earth, water, fire) and set in the four seasons (spring, summer, autumn, winter). In addition, the poem overall deals with the four basic concepts of time: time present, time past, time future, and timelessness.

secondary cell a cell in which the chemical action is reversible.

secondary colour a colour formed by mixing two **primary colours**.

secondary growth cancer elsewhere in the body than on the original site.

secondary modern school a **secondary school** offering a more practical or a less academic education than a grammar school.

secondary picketing picketing by workers of a firm with which they are not directly in dispute but with which they have a trading or other connection.

secondary school a school providing education for children between 11 and 16 (or 18), to school-leaving age.

secondary sexual characteristics characteristics that indicate a person's or animal's sex but that are not involved in reproduction, such as a man's beard or a deer's antlers.

second ballot a ballot between the two candidates who received the highest vote in an earlier election but who failed to win an absolute majority, the candidate scoring the lower number of votes being eliminated.

second banana a person playing an inferior role or doing something less important than another (originally a music-hall performer in a subsidiary role).

second best the next best alternative, not the first choice.

second childhood dotage, senility (old people sometimes become childish or childlike).

second class a service that is inferior in some way to one that is **first class**, such as a less comfortable seat on public transport, or a slower rate of mail delivery (the second class of rail travel is now called 'standard class' by British Rail).

Second Coalition the alliance of Britain, Austria and Russia to drive the French out of Italy, Switzerland and the German states in 1799.

Second Coming the prophesied return of Christ at the Last Judgement.

second course the sweet course or pudding of a main meal.

second cousin the child of a **first cousin** of one of a person's parents.

Second Covenant God's promise to Israel that he would forgive their sins (*Jeremiah* 31:31–4, *Hebrews* 8:8–13), this promise being fulfilled in the coming of Jesus.

Second Day among Quakers, the (former) name of Monday.

second degree burn a burn in which blisters appear on the skin.

Second Empire the imperial government of France under Napoleon III from 1852 to 1870, also the ornate furniture typical of this period.

second estate the nobility of a country.

Second Eve a name for the Virgin Mary, as the equivalent of Eve, the first woman and mother (compare **Second Adam**).

second fiddle the player of a second violin part in an orchestra or string quartet, and so a term generally for a less important person.

second floor the floor two storeys up from the ground (in UK) or one storey up (US) (compare **first floor**).

Second Folio the second published collection of Shakespeare's plays (1632).

Second Front the attack planned on Europe by the Western Allies after the fall of France in 1940.

second generation (1) the offspring of immigrant parents; (2) an improved version of a machine or model (of computers, those built between the mid-1950s and mid-1960s, using transistors instead of valves).

second-hand not new, used or previously owned (compare **first-hand**).

second home a holiday home, whether for weekend use or for longer periods.

Second Inaugural Address US President Lincoln's address at the commencement of his second term of office (1865), with its famous words 'with malice toward none; with charity for all'.

second-in-command (2 i/c) the assistant to a commanding officer (compare **number two** (2)).

Second International the international association of socialists and trade unionists that met in Paris in 1889 (compare **Second-and-a-half International**, **Third International**).

second law of motion the most important of the three laws, stating that the rate of change of momentum is directly proportional to the force, taking place in the direction of the force.

second law of thermodynamics the law that it is impossible for any self-acting machine, without any other agency, to transfer heat to another body at a higher temperature; in other words, heat always flows from a higher temperature to a lower.

second lieutenant an army officer of the lowest commissioned rank, or his equivalent in the Royal Marines, the US Army, USAF, and US Marines.

second man the assistant to a locomotive driver (previously called the fireman).

second mate the next in command of a merchant vessel after the **first mate**.

second mortgage a mortgage arranged after the first, usually to help with payments.

second name surname, family name, last name.

second nature a habit that is longstanding and easily put into practice, but not one that is innate.

second person in grammar, the person spoken to, usually expressed by 'you' (formerly also 'thou' and 'thee').

second position in ballet, the stance with feet wide-splayed and one step apart, the toes pointing left and right, and the accompanying pose with arms apart at shoulder height.

second-rate inferior.

second reading the second presentation of a bill in Parliament, to be generally approved in principle (or, in US, to have the committee's report on it discussed).

Second Reich the German Empire from 1817 to 1918.

Second Republic the republican government of France from the deposition of Louis-Philippe (1848) to the **Second Empire** (1852).

seconds another helping of food.

second self a person closely identified with oneself.

second sight clairvoyance, the ability to see future events or events happening elsewhere.

second strike in nuclear warfare, a counter-

attack following an initial enemy attack (compare **first strike**).

second string (1) an alternative course of action, a 'second string to one's bow' (as formerly carried by archers); (2) a substitute or reserve player in a sport; (3) an inferior person.

second thoughts a change of mind, a revised opinion.

second to none supreme, unsurpassed.

second wind renewed strength after the expenditure of energy.

Second World War the global conflict of 1939 to 1945 (compare **First World War**).

BASED ON **DOUBLE**

double (1) a double measure of spirits; (2) a person or ghost that is the exact counterpart of someone; (3) a stroke in billiards in which the ball rebounds off a cushion to enter a pocket; (4) a bet on two horses; (5) the narrow outer ring on a dartboard, or a hit on it.

double agent a spy working for two countries that are mutually hostile.

double-barrelled (1) a gun that has two barrels; (2) a surname that consists of two words, often hyphenated.

double bass a stringed instrument with a large, low range of notes (so named by contrast with the 'single' cello; a string twice the length of another gives a note an octave lower).

double bassoon the lowest-noted and largest instrument in the oboe class (so named by comparison with the 'single' bassoon).

double bed a bed for two people.

double bill an entertainment with two main items.

double-booked a seat or room that has been reserved twice to ensure its occupancy in case of cancellation.

double-breasted a coat or jacket with overlapping fronts, giving a double layer of cloth, and having a double row of buttons or fasteners, but only a single row of buttonholes.

double check to check something twice, especially as a precaution.

double chin a chin with a fold of fat under it.

double concerto a concerto for two instruments.

double cream thick cream, with a high fat content.

double-cross to cheat, deceive (compare **two-time**).

double date in US, a 'date' in which two couples go out together.

double dealing a treacherous or deceitful act, duplicity.

double-decker a bus with two decks, upstairs and downstairs, or any object arranged in two decks or layers.

double dipper in US, a government employee who draws a pension from one department while working for another.

Double Double the nickname of the star *Epsilon Lyrae*, which is a quadruple, that is, two combined **double stars**.

double Dutch gibberish, incomprehensible talk.

double-edged able to be understood in two ways, having two implications.

double entendre a remark that has an indelicate meaning as well as a standard one (from obsolete French, meaning 'double meaning').

double entry a bookkeeping system with an entry showing the debit in one column and the credit in another.

double exposure a photograph with two superimposed images, made accidentally or deliberately.

double fault in tennis, the serving of two faults (such as two balls into the net), so losing the point.

double first a Class 1 honours degree in two subjects.

double flat in music, an accidental (symbol *bb*) that lowers the pitch of the note two semitones.

double glazing two panes of glass in a window to reduce loss of heat and the transmission of outside noise (compare **triple glazing**).

double helix the molecular structure of DNA, which consists of two helical chains coiled round the same axis.

double indemnity an insurance policy that pays double the face value in case of accidental death.

double-jointed having unusually flexible joints, as if two.

double life a deceitful life.

double or quits a move in a card game to decide whether a stake is to be doubled or cancelled.

double-parked of a vehicle parked next to another by the roadside, so causing an obstruction.

double pneumonia pneumonia in both lungs.

double room a room for two people, especially a bedroom with a **double bed** or twin beds.

double saucepan a single cooking utensil consisting of two saucepans one inside the other, with the water boiling in the bottom pan to heat the food in the upper one.

double sculls a racing shell with two scullers sitting one behind the other, each having two oars.

double sharp in music, an accidental (symbol **𝄪**) that raises the pitch of the note two semitones.

double-space to type with a full space between the lines.

double standard a set of principles that allows greater freedom to one person or group than to another.

double star two stars that are either linked closely to one another by gravity or that lie on the same line of sight from the Earth, so seeming to be linked (but actually being far apart).

double take a second, delayed reaction to something, often used for comic effect.

double talk 'hot air', words that seem to have a grand or important meaning but that are really commonplace and not specially significant.

double think the deliberate or unconscious interpretation of conflicting facts.

double time (1) in US Army, a fast march of 180 paces to the minute; (2) double the rate of normal pay, as for overtime or work done on a holiday.

doubleton two playing cards only of a suit (in bridge) (compare **singleton**).

double top in darts, a score of double 20 (which is at the top of the board).

double up (1) to share accommodation or facilities with someone else; (2) to bend the body in two when laughing or in great pain.

BASED ON **TWIN**

twin bed one of a matching pair of **single beds**.

twin city in North America, one of two cities that are close together, such as St Paul and Minneapolis (USA) or Fort William and Port Arthur (Canada).

twin double in horseracing, a system of betting in which the winners of 4 successive races must be forecast, i.e. two **doubles** (4) in sequence.

twin jet an aircraft with two jet engines.

twin lens of a camera, having two identical sets of lenses, one for viewing, and one for exposing the film.

twin paradox in physics, the theory that if one of a pair of twins makes a long journey at high speed and then returns, he will have aged less than the twin who stays behind.

twin-plate glass glass that is ground and polished on both sides at once.

twin screw having two propellers on separate shafts.

twin-set a woman's matching jumper and cardigan.

twin town a town in one country officially linked with another town in another country, with which it exchanges cultural visits and the like; the two towns are usually similar in size or commercial activity.

twin-tub a washing machine with two revolving drums, one for washing and the other for spin-drying.

QUOTATIONS AND SAYINGS

A bicycle made for two. (Harry Dacre, song 'Daisy Bell', *c.* 1892)

Can two walk together, except they be agreed? (Bible: *Amos* 3:3)

In two words: im-possible. (Samuel Goldwyn, quoted by Alva Johnson, *The Great Goldwyn*)

It takes two to tango. (Hoffman and Manning song title, 'Takes Two to Tango', 1952)

Mankind is divisible into two great classes: hosts and guests. (Max Beerbohm, *Hosts and Guests*)

No man can serve two masters. (Bible: *Matthew* 6:24)

Tea for two, and two for tea. (Otto Harbach and Frank Mandel, song title from musical *No, No, Nanette*, 1925)

That makes two of us. (saying)

There are two sides to every question. (Protagoras, 5th century BC)

Two can play at that game. (saying)

Two heads are better than one. (proverb)

Two little girls in blue. (Vincent M. Youmans and Paul Lannin, title of musical, 1921)

Two lovely black eyes. (Charles Coborn, song title)

Two red roses across the moon. (William Morris, poem title)

Two's company, three's none. (proverb)

Two wrongs don't make a right. (saying)

What is a friend? A single soul dwelling in two bodies. (Aristotle, 4th century BC)

THREE

EXPRESSIONS

BASED ON THREE

Battle of the Three Emperors the Battle of Austerlitz (1805), when the French emperor Napoleon routed the emperors of Austria (Francis I) and Russia (Alexander I), all three being personally present on the battlefield.

Big Three (1) Clemenceau, Lloyd George, Woodrow Wilson, representing respectively France, Britain and US at the Paris Peace Conference of 1919; (2) Churchill, Franklin D. Roosevelt, Stalin, the respective heads of Britain, USA and USSR in the Second World War, who met at Yalta in 1945 to discuss the final defeat of the Nazis; (3) the athletic teams of Harvard, Princeton, and Yale Universities, USA.

Charlotte Three James Earl Grant, Jr, Charles Parker, Thomas J. Reddy: three black activists jailed in 1972 for the burning of a whites-only stable in Charlotte, North Carolina.

League of the Three Emperors the Dreikaiserbund of 1872, whereby the emperors of Germany (William I), Austria (Francis Joseph I) and Russia (Alexander II) agreed to co-operate in maintaining a status quo in Europe.

ménage à trois a household of two women and one man, or two men and one woman (French, 'household of three').

page three Page 3 of *The Sun* newspaper, on which a photo of a nude or near-nude female appears daily.

rule of three the mathematical rule stating that the value of one unknown quantity in a proportion is found by multiplying the denominator of each ratio by the numerator of the other, that is, the 1st is to the 2nd as the 3rd is to the unknown 4th, for example, $9 \div 3 = 18 \div x$; $(3 \times 18) \div 9 = x$; so $x = 6$, otherwise 3 is one third of 9, so x is one third of 18.

three ages in archaeology, the Stone, Bronze and Iron Ages.

three As a short name for the Amateur Athletics Association.

Three Bishoprics the bishoprics of Metz, Toul and Verdun, which governed ancient France, and which were seized by Henri II in 1552.

three-card trick the trick also known as 'find the lady', in which bets are placed on which of three inverted cards is the queen.

three cheers the traditional shouts of 'hurrah' in unison, each usually expressed as 'hip, hip, hip' (called by one person) 'hooray' (shouted by all) (compare **two cheers**).

Three Choirs Festival the choral festival held

DANTE'S MYSTIC 3

Dante's great epic poem, *The Divine Comedy*, contains many direct and indirect references to the mystic number 3.

The poem is written in *terza rima* and its 99 cantos, after an introductory canto, are divided equally into three sections of 33 cantos each. The sections deal respectively with the poet's journey to Hell, Purgatory and Paradise, and the word 'three' itself occurs several times in the work.

On beginning his journey into Hell, the poet is pursued by three beasts, a panther, a wolf and a lion. Later, he becomes convinced that three blessed women are anxious about his fate. On reaching the centre of the earth, the poet finds Lucifer with three heads gnawing on three infamous traitors: Brutus and Cassius, who betrayed Julius Caesar, and Judas Iscariot, who betrayed Jesus.

annually since 1724 at Gloucester, Worcester and Hereford, with each cathedral rotating in turn to accommodate the three choirs from each.

three-colour process the process that gives a picture by superimposing three prints in inks that correspond to the three **primary colours.**

three-cornered fight a parliamentary contest with three candidates, in the late 1980s usually Conservative, Labour and SLD (earlier, Liberal).

three-day event an equestrian contest lasting three days, usually with dressage on the first day, an endurance ride, cross-country race and steeplechase on the second, and jumping (in a ring) on the third.

three-decker a sandwich with three slices of bread and two fillings.

three-dimensional (3-D) of a film, having the effect of depth by presenting slightly different views of a scene to each eye (compare **third dimension**).

three-dog night a very cold night (needing three sled dogs to keep warm when bedding down for the night in the Arctic).

three-eye league a hypothetical club to which a politician belongs who has visited the three homelands of the US's major minorities: Israel, Italy and Ireland (all beginning with 'I').

three-field system a system of crop rotation practised in England until the 18th century, with three open arable fields used successively for wheat or rye, peas, beans, barley or oats, and left fallow for the third season.

three-fifths compromise in the US, a system whereby in order to redress the imbalance of representation between the populous North and the sparsely settled South, the southern states were allowed to count each slave as three-fifths of a person for their congressional appointment, slaves having originally been regarded as property, and having no vote; the system meant that the more slaves there were, the less power they had (and the more power the slaveowners had).

three-four time in music, a measure of three crotchets (quarter notes) to a bar, indicated by the time signature $\frac{3}{4}$.

three-gaited of a horse, having three paces: walk, trot and canter.

three golden balls a pawnbroker's sign, with the three balls adopted from the coat of arms of the Medici family, and introduced to London by the Lombard bankers; the positioning of the balls, 1 under 2, was popularly supposed to represent a 2-to-1 chance of redeeming what was pawned.

three-halfpence 1½d, in pre-decimal money.

three heart rule in the US, the rule that any member of the US forces wounded three times in the same tour of duty, so winning three Purple Hearts, is evacuated from the combat zone.

Three Holy Children Shadrach, Meshach and Abednego (in the Benedicite: Ananias, Azarias and Misael), who praised God with Daniel in King Nebuchadnezzar's 'burning fiery furnace' (*Daniel* 3).

three hours service a religious service held on Good Friday from 12.00 noon to 3 p.m., with seven sermons on the **Seven Words from the Cross**.

Three in One (One in Three) a term used of the Holy Trinity.

Three Jesuit Martyrs Edmund Campion, Alexander Briant and Ralph Sherwin, who were executed at Tyburn in 1581 but who were canonized with the **Forty Martyrs** in 1970.

Three Kings Day another name for the Epiphany, when the **Three Wise Men** came from the east to the infant Jesus.

three-legged race a race in which competitors run with one leg tied to that of another person, making a 'single' leg which has to be synchronized with the two outer legs.

three-line whip in Parliament, an instruction to MPs to attend and vote without fail (the instruction being underlined three times) (compare **two-line whip**).

three martini lunch in US, a nickname for a lavish business lunch on expenses, especially one that is unduly lengthy, over-indulgent, and unproductive.

three-mile limit in international law, the range of a nation's territorial waters, extending to three nautical miles from the shore (originally said to be the range of a smooth-bore cannon).

three of a kind three identical objects, especially three playing cards of the same suit or value, but also three similar people.

threepenny bit a former British 12-sided coin, value 3d, obsolete from 1971.

three-piece of a suit or costume, consisting of

DANTE'S MYSTIC 9

Dante's great poem, *The Divine Comedy*, contains almost as many references to the mystic number 9 as it does to 3.

There are nine descending levels of Hell in Part 1, nine ascending steps up the mountain in Purgatory in Part 2, and nine heavenly spheres in Paradise in Part 3.

The figure 9 was also connected with Dante's personal life. He first met his beloved Beatrice when he was 9, and met her for the second time 9 years later, when he dedicated his first sonnet to her. Beatrice died in 1290, the year that closed the 9th decade of the century.

three pieces, such as a suit of jacket, trousers and waistcoat, also a matching furniture suite of two armchairs and sofa (compare **one-piece**, **two-piece**).

three pipe problem a problem that requires much thought (as if smoking three pipes successively).

three-point landing of an aircraft, with the two main wheels and the nose (or tail) touching down simultaneously; regarded as the perfect landing.

three-point turn of a motor vehicle, reversing the direction of motion by driving forwards in one turn, then backwards in an opposite turn, then forwards again in a turn opposite to the first.

three-quarter a rugby player between the full back and the forwards.

three-ring circus a circus with three rings in which three performances go on simultaneously, hence any scene of confused activity.

three Rs the fundamentals of education: reading, writing, and arithmetic (that is, reading, 'riting, 'rithmetic) (compare quotation by Tom Pollock under 'Four').

three sheets in the wind very drunk (that is, like a ship with three sails flapping in the wind because their sheets [attaching ropes] are not fastened).

threesome a group of three players, especially in golf, where one player, with his own ball, plays against two players who hit a single ball alternately.

three-speed gear a gear, usually on a bicycle, with three speeds: low (for uphill), medium (for flat) and high (for downhill), the legs pedalling at approximately the same rate to achieve the appropriate speed.

three-square of a file, having a cross-section that is an equilateral triangle.

Three Stooges Larry Fine, Moe Howard, Jerry Howard, three well-known US comedians.

three times three **three cheers** repeated three times.

Three Tongues the languages in which the inscription on Jesus's cross was written at the Crucifixion: Hebrew, Greek and Latin (languages regarded as essential knowledge for medieval theological students, and in which the earliest versions of the Bible were written).

three-wheeler a light motor vehicle with three wheels, corresponding to a tricycle.

Three Wise Men the Magi, traditionally named as Melchior, Gaspar (Caspar) and Balthazar, and the three kings from the east who brought three gifts to the infant Jesus; the gifts were respectively gold (symbolizing royalty), frankincense (divinity) and myrrh (prophesying the bitterness of persecution and death).

three wishes the traditional number of wishes (from an old fairy story, told in Persia in the 7th century).

three worlds the concept of the universe in Hindu mythology, consisting of earth, middle space (atmosphere), and ether (sky); or men, semi-divine creatures and gods; or heaven, earth and the territory of the demons (hell).

BASED ON **THIRD**

third class one of the lowest standards of anything, such as an academic degree, a public transport service, or a rate of mail delivery (as in the US for printed matter below a certain weight).

Third Coalition the alliance of Britain, Russia, Sweden and Austria against Napoleonic France and Spain in 1805 (third by comparison with the first [Austria, Prussia, Spain, United Provinces, Britain] of 1793 and second [Britain, Russia, Ottoman Empire, Naples, Portugal] of 1798.

Third Day the (former) Quaker name for Tuesday.

third degree the use of torture or bullying to extract confessions or information (named after the highest degree of Masonic ritual).

third degree burn the destruction of the epidermis and dermis by burning, that is, the death of all layers of skin.

Third Department the former Russian political

security department, created 1826 and abolished 1880 (officially the Third Section of His Imperial Majesty's Own Chancery in Russia; Russian: *tret'ye otdeleniye*).

third dimension the dimension that enables a solid object to be distinguished (in depth) from a two-dimensional picture or drawing of it (compare **three-dimensional**).

third estate the commons, townsmen or middle classes of a country.

third eye in Buddhism, the eye of insight or destruction, located in the middle of the forehead of the god Siva, hence generally a synonym for 'intuition', 'inner sight'.

third eyelid the nictitating membrane in many reptiles and birds, being the fold of skin below the eye that can be drawn across the eye.

third force a force standing between two political groups or parties, as originally between the French Gaullists and the Communists (French: *Troisième Force*), but now, any force standing between the two superpowers (USA and USSR).

third generation (1) the offspring of immigrant grandparents; (2) an improved version of nuclear weapon or computer (of the latter, one produced between 1966 and 1979, in which integrated circuits replaced transistors).

third house in US, a political lobby for a particular interest (by contrast with the two main parties, Republicans and Democrats).

Third International the Comintern (Communist International), an international Communist organization founded by Lenin in Moscow in 1919 (dissolved in 1943) (compare **Second International**, **Second-and-a-half International**).

third law of motion the law that the actions of two bodies on each other are always equal and opposite, that is, that a reaction is always equal and opposite to an action.

third law of thermodynamics the law that the entropy (measure of disorder of molecules and atoms) of disordered solids reaches zero at absolute zero of temperature, that is, that there will then be no disorder at all.

third man in cricket, a fielding position (or fielder) on the offside near the boundary behind the batsman's wicket (3rd by comparison with point and slip).

third man argument in the philosophy of Aristotle, the man who seems to be needed in order to argue from the particular instance of a man to the ideal form of a Man (Greek: *tritos anthropos*).

third market the trading of stocks and shares outside the stock exchange or the unlisted securities market.

third order the religious society of lay people affiliated to a religious order but observing only a modified form of religious life (Latin *tertius ordo*, and regarded as third by contrast with the first order of monks and the second order of nuns; also known as tertiary order).

third party insurance insurance that provides protection against liability to a third person by two principals, the insured and the insurer.

third person in grammar, the person(s) or thing(s) spoken about by two other persons (**first person** and **second person**), either by noun or name or as a pronoun ('he', 'she', 'it', 'they', etc.).

third position in ballet, a stance with the feet wide-splayed, one before the other, with the heel of the front foot fitting into the hollow of the instep of the back foot, while the arms are raised over the head.

Third Programme the former music and 'cultural' radio programme of the BBC, from 1946 to 1967, when it became Radio 3.

third rail the rail beside the two main rails from which an electric train picks up its current.

third-rate inferior, of poor quality.

third reading in UK, the parliamentary process of discussing the committee's report on a bill, or (in US) the final discussion of a bill before a vote is taken on its implementation.

Third Reich the Nazi dictatorship in Germany from 1933, when Hitler came to power, to his death in 1945, at the end of the Second World War (compare **Thousand-Year Reich**).

Third Republic the French government from the fall of Napoleon in the Franco-Prussian War (1870) to the German occupation (1940).

Third Rome Moscow, so named in the 16th century by influential writers after the fall of the Roman Empire (1st Rome) and Constantinople (2nd Rome), the latter in 1453.

third sex homosexuals, regarded as a distinct sexual group.

third stream in music, a style combining jazz and classical music (third after classical music itself [first] and jazz [second]).

Third World the countries of Africa, Asia and

Latin America, regarded as underdeveloped and neutral by contrast with the two superpowers (USA and USSR) (with the 'Second World' being either the developed countries apart from the superpowers, or the Communist bloc).

Third World War a hypothetical future global conflict.

BASED ON **TREBLE** AND **TRIPLE**

treble bob in bellringing, a form of change-ringing with the treble having a uniform but zigzag ('bobbing') course, and with all bells dodging.

treble chance in football pools, a bet with chances of winning related to the number of draws and of home and away wins forecast.

treble clef (𝄞) in music, the clef for the upper notes, centred on G, a fifth above middle C.

treble stringing in tennis, the fine gut binding at the top and bottom of the racket head, looping three times round each main gut string.

Triple Alliance the alliance (1) between Germany, Austria-Hungary and Italy from 1882 to 1914; (2) between France, the Netherlands and Britain against Spain in 1717; (3) between England, Sweden and the Netherlands against France in 1668.

triple bars in equestrianism, an obstacle consisting of three bars placed in sequence, but descending in height, technically a type of spread fence.

Triple Crown (1) the notional title of the horse that wins the three Classic races (in UK: **Two Thousand Guineas**, Derby and St Leger; in US: Belmont Stakes, Kentucky Derby and Preakness Stakes); (2) the similar title of the rugby team that in one season defeats all three opponents, the teams being those of England, Scotland, Wales and Ireland.

triple double in baseball, a player who in a single game gets double figures in scoring, rebounds and assists.

Triple Entente the understanding between Britain, France and Russia from 1894 to 1907 to counterbalance the **Triple Alliance** (1); it formally ended in 1917.

triple glazing glazing with three panes of glass (compare **double glazing**).

triple jump in athletics, the hop, step and jump.

triple play in baseball, play with three men put out before the ball is returned to the pitcher for delivery.

triple time in music, a measure of three beats in a bar.

triple-witching hour in US, the first hour of trading on the New York stock exchange on four annual days (third Friday of March, June, September and December) when contracts expire simultaneously on (a) futures, (b) options on indexes, and (c) options on various individual stocks.

JULIET 'COMES OF AGE'

In Shakespeare's *Romeo and Juliet*, Juliet's mother, Lady Capulet, discusses her daughter's age and marriage prospects with Juliet's nurse, having summoned Juliet herself to be present.

Lady Capulet [to Nurse]: . . . Thou know'st my daughter's of a pretty age.
Nurse: Faith, I can tell her age unto an hour.
Lady Capulet: She's not fourteen.
Nurse: I'll lay fourteen of my teeth—
And yet to my teen [grief] be it spoken I have but four—
She is not fourteen. How long is it now
To Lammas-tide?
Lady Capulet: A fortnight and odd days.
Nurse: Even or odd, of all days in the year,
Come Lammas-eve at night shall she be fourteen.

Lammas is 1 August. Juliet's birthday was thus on the eve of this, 31 July. Hence her name. And Lady Capulet regards her daughter, at nearly 14, to be old enough to marry the young nobleman, Paris. But events turned out otherwise . . .

QUOTATIONS AND SAYINGS

All really great men may be said to live three lives: there is one life which is seen and accepted by the world at large, a man's outward life; there is a second life which is seen by a man's most intimate friends, his household life; and there is a third life seen only by the man himself and by Him who searcheth the heart. (Max Müller, quoted in *The Cabinet Portrait Gallery*, 3rd series, 1892)

And now abideth faith, hope, charity, these three; but the greatest of these is charity. (Bible: *I Corinthians*, 13:11)

Baa, baa, black sheep,
Have you any wool?
Yes, sir, yes, sir,
Three bags full (nursery rhyme)

Did you never see the picture of We Three? (Shakespeare, *Twelfth Night*, 1601) (the reference is not to the characters present, Sir Andrew Aguecheek, Sir Toby Belch and the Clown, but to an inn sign, 'The Two Loggerheads', with the inscription 'we three loggerheads be', the third being the spectator)

Gallia est omnia divisa in partes tres (Gaul as a whole is divided into three parts). (Caesar, *De Bello Gallico*, c. 58–44 BC)

He that wants money, means, and content is without three good friends. (Shakespeare, *As You Like It*, 1599)

I saw three ships come sailing in. (traditional Christmas carol)

It is by the goodness of God that in our country we have those three unspeakably precious things: freedom of speech, freedom of conscience, and the prudence never to practise either of them. (Mark Twain, *Following the Equator*, 1897)

No guest is so welcome in a friend's house that he will not become a nuisance after three days. (Titus Maccius Plautus, *Miles Gloriosus*, c. 205 BC)

One hour's sleep before midnight is worth three after. (George Herbert, *Jacula Prudentum*, 1651)

Self-reverence, self-knowledge, self-control,
These three alone lead life to sovereign power.
(Tennyson, *Oenone*, 1832)

The antique Persians taught three useful things,
To draw the bow, to ride, and speak the truth.
(Byron, *Don Juan*, 1819–24)

There are three kinds of lies; lies, damned lies and statistics. (Disraeli, *Autobiography*, attributed to Mark Twain)

There are three things which the public will always clamour for, sooner or later: namely, Novelty, novelty, novelty. (Thomas Hood, announcement of *Comic Annual* for 1836)

The three great elements of modern civilization, Gunpowder, Printing, and the Protestant Religion. (Thomas Carlyle, *State of German Literature*)

The three most beautiful things in the world: a full-rigged ship, a woman with child, and a full moon. (anonymous)

Third time lucky. (popular saying, of ancient origin)

This is the third time of asking. (Book of Common Prayer, wording of banns of marriage, which 'must be published in the Church three several Sundays')

Three acres and a cow. (Jesse Collings, 1880s; Collings, a henchman of the British prime minister Joseph Chamberlain, proposed that every smallholder in Britain should have these things; the phrase itself originated in John Stuart Mill's *Principles of Political Economy*, 1848: 'When the land is cultivated entirely by the spade and no horses are kept, a cow is kept for every three acres of land'; on advocating this policy, Collings was nicknamed 'Three Acres and a Cow Collings'; later, Noël Coward would nickname Edith, Osbert and Sacheverell Sitwell as 'two wiseacres and a cow'; compare 'ten acres and a mule', p. 119 and 'forty acres and a mule', p. 130)

Three blind mice. (nursery rhyme, said by some to refer to Archbishop Cranmer, Bishop Ridley and Bishop Latimer, the three Protestant churchmen who were burnt at the stake in 1555–56 by the Catholic queen, Mary Tudor, 'the farmer's wife')

Three little maids from school are we. (Gilbert and Sullivan, *The Mikado*, 1885; the maids are Yum-Yum, Pitti-Sing, and Peep-Bo)

Three may keep a secret, if two of them are dead. (Benjamin Franklin, *Poor Richard's Almanac*, 1733–58)

Three o'clock is always too late or too early for anything you want to do. (Sartre, *Nausea*, 1938)

When shall we three meet again? (Shakespeare, *Macbeth*, 1606; the famous opening line refers to the Three Witches)

FOUR

EXPRESSIONS

BASED ON **FOUR**

Annals of the Four Masters a collection of 16th-century Irish chronicles, whose compilers where Michael O'Clery, Conaire O'Clery, Cucoigriche O'Clery, Fearfeasa O'Mulconry.

Big Four (1) Britain, France, the USSR and the USA after the Second World War, as four influential countries; (2) in UK, the four main banks:

National Westminster, Midland, Lloyds and Barclays (which before 1968 were in the **Big Five**); (3) the constructors of the first transcontinental railroad in the US, the Central Pacific and Union Pacific Railroads: Charles Crocker, Collis P. Huntington, Mark Hopkins, Leland Stanford.

Clause 4 the Clause in the Labour Party's manifesto that declared its commitment to public ownership in the 1980s.

Fab Four the Beatles: John Lennon, Paul McCartney, George Harrison, Ringo Starr.

fire on all four cylinders to be in perfect condition, like a smoothly-running engine.

four-ale bar a public bar in a pub, regarded (formerly) as a place of cheap enjoyment ('four ale' was cheap mild ale, originally sold at 4d a quart).

four ball a golf match for two pairs of players, with each player using his own ball, and the better score of each pair counted at each hole.

four-by-two (1) a piece of timber with a cross-section measuring 4 × 2 in; (2) in army slang, a piece of cloth this size attached to a pull-through to clean a rifle.

four-colour printing printing using red, blue, yellow and black in combination to produce almost any other colour.

four-colour problem the problem, unsolved until 1976, of proving that only four colours are needed on a map to give different colours to adjacent or adjoining territories.

four corners of the earth everywhere, worldwide.

Four Courts Dublin's central courts of justice (now actually ten courts).

Four Crowned Martyrs the *Quattuor Coronati*, or two groups of 4th-century martyrs, one actually of five Persian stonemasons, the other of four Roman soldiers, and both probably mutually identified; their names are usually given as Claudius, Nicostratus, Simpronian and Castorius (with the fifth, who later dropped out of the legend, as Simplicius).

four-dimensional involving the three spatial dimensions and the fourth dimension of time (see also **fourth dimension**).

four eyes a disparaging nickname for a person wearing spectacles.

4-F in US, a person rejected for military service because of physical, mental or moral instability (from the low arbitrary rating used in the Second World War; compare **A1**).

four-five-six a dice game, also known as see-low, played with three dice in which 4-5-6 is one of the winning combinations.

four flush a useless hand at poker, containing four cards of a suit and one odd one (instead of the required five).

four-footed friend a nickname for a pet animal, such as a dog or cat.

four-foot way in the UK, the standard railway gauge (actually 4 ft 8½ in (1435 mm) between rails) (compare **six-foot way**).

four-four time in music, a quadruple time with crotchets (quarter notes) to the bar, denoted as $\frac{4}{4}$.

Four Freedoms the basic human freedoms of speech, religion, from want and from fear, as enunciated by Franklin D. Roosevelt in 1941.

Four-H Club in US, a club in a nationwide organization of young people in rural areas, encouraging modern farm practices and good citizenship (so named with the aim of improving Head, Heart, Hands and Health).

Four Horsemen the name given by the US sportswriter Grantland Rice, of the *New York Herald Tribune*, to the backfield of Notre Dame's undefeated football team of 1924: Harry Stuhldreher (quarterback), Don Miller and Jim Crowley (halfbacks), Elmer Layden (fullback); the four were photographed on horseback wearing football uniforms, which caught the fancy of their fans.

Four Horsemen of the Apocalypse the personification of war, famine, pestilence and death as four horsemen in the Bible (*Revelation* 6:2–8).

four-in-hand a road vehicle drawn by four horses and driven by one driver (who has the reins of all four in his hand), also, any four-horse team in a coach or carriage, whether literally 'in hand' or not.

Four Last Things in theology: death, judgment, heaven and hell, the study of which is known as eschatology.

four-leaf clover a rare clover with four leaves instead of the usual three, and popularly regarded as a sign of good luck.

four-letter word a short English word referring to sex or excrement, and often used as a swear word; there are hardly more than a 'dirty dozen' of such words, and only two or three which are still regarded as taboo or very offensive.

four-minute man in US, a man who in the First World War made a short speech to promote the sale of government bonds.

four-minute mile the running of a mile in 4 minutes or less, first achieved (in 3 min 59.4 s) on 6 May 1954 by the British athlete Roger Bannister.

Four Noble Truths in Buddhism, the truths that: life is suffering; the cause of suffering is 'birth sin', or craving and desire; suffering can be ended only by Nirvana, which is the extinguishment of desire; Nirvana itself is accomplished by the **Eightfold Path** to righteousness.

four of a kind four cards of a set or suit, or four identical or similar objects, or people in general.

four on the floor a standard gear lever on the floor of a car, where it controls the car's four speeds (as distinct from an automatic gear change).

four-pack a pack of four identical articles sold together, such as four cans of beer.

fourpenny one a blow with the fist (probably rhyming slang in origin, with 'fourpenny bit' = 'hit').

four poster a bed with posts at each corner supporting a canopy and curtains.

Four-Power Pacific Treaty the treaty of 1921 on limitation of armament, between the USA, UK, France and Japan.

four seas the seas round Britain: Western (Irish Sea), Northern (North Sea), Eastern (former German Ocean) and Southern (English Channel).

four seasons spring, summer, autumn, winter, often used as a collective name, for example of a restaurant (as the Four Seasons, New York).

four senses (four levels of meaning) the four ways of interpreting allegorical literature, and especially the Bible: (1) historical or literal, (2) allegorical, (3) moral, (4) anagogical (mystic); for example, Jerusalem is (1) a city in Palestine, (2) the Church, (3) the believing soul, (4) heaven.

four signs in Buddhism, the four encounters with old age, sickness, death and austerity which in legend convinced the young prince Siddhartha that he should renounce the royal life and strive for Buddhahood.

foursome a group of four players, teamed as two opposing pairs, for example, in tennis.

foursquare firm(ly), secure(ly), fair(ly) (as if square-shaped, and so solid and dependable).

four-stroke engine an internal combustion engine with the piston making four strokes every explosion (intake, compression, combustion, exhaust).

four-wheel drive a motor vehicle system with all four wheels connected to the source of power, not simply the back two, as is most common.

Gang of Four (1) in China, the leaders of the Cultural Revolution who were accused of counter-revolutionary revisionism after the death of Chairman Mao in 1976 and who were tried four years later: Wang Hung-Wen, Chang Ch'un Chiao, Chiang Ch'ing (Jiang Qing) (Mao's widow), Yao Wen-Yuan; (2) the Labour MPs who broke away in 1981 to found the SDP: Roy Jenkins, Shirley Williams, David Owen, William Rodgers.

Guildford Four the four alleged IRA terrorists who planted bombs in Guildford pubs in 1974 and who were jailed for life: Paul Hill, Gerard Conlon, Patrick Armstrong, Carole Richardson; others, however, later claimed the 'credit' for the bombings.

on all fours on hands and knees, or hands and feet.

plus-fours baggy trousers with a fold below the knee, traditionally worn by golfers (named from the extra 4 inches of cloth needed).

Point-Four Program in US, the programme for aid to the **Third World** made in 1947 as the fourth point of President Truman's inaugural address.

within these four walls between you and me, in confidence.

BASED ON **FOURTH**

fourth class the lowest category of something, such as an academic degree (formerly at Oxford) or mail service (in US, the latter used for parcels and merchandise).

Fourth Day the (former) Quaker name for Wednesday.

fourth dimension (1) time, especially as necessary in addition to the three spatial dimensions to specify the position and behaviour of a point or particle; (2) in science fiction, a dimension additional to the other three used to explain supernatural phenomena and the like.

fourth estate the press, journalists.

THE THIRD WHAT?

The Latin term *tertium quid* means literally 'third something', and is used to refer to an unknown or indefinite thing that is in some way related to two known or certain things, while at the same time being distinct from them. For example, the 19th-century English philosopher James Martineau described God as a *tertium quid*, 'merely an external relation', to mind and matter. But, more commonly, *tertium quid* is used simply of an unknown third person or thing, as when Rudyard Kipling began a short story ('At the Pit's Mouth') with the words: 'Once upon a time there was a Man and his Wife and a *Tertium Quid*'. The Latin expression was itself a translation of Greek *triton ti*, 'some third thing'.

fourth generation of a computer, built during the 1980s and the most sophisticated before the advent of even more advanced **fifth generation** computers.

Fourth International the Trotskyist organization first formed in 1938 as an anti-Stalinist grouping, condoning Fascism, and designed to attract Communists away from the Comintern (see **Third International**).

fourth leader the humorous or light-hearted leading article that appeared under the three main (serious) leaders in *The Times* from 1922 to 1966; it was known by the newspaper's staff as 'the funny'.

fourth market the direct trading of unlisted securities on the stock market, thus bypassing the conventional method (compare **third market**).

Fourth of July in US, Independence Day, the annual public holiday celebrating the day of adoption of the Declaration of Independence on 4 July 1776.

Fourth of June the annual speech day and celebrations held at Eton College to mark the birthday of one of the school's most famous patrons, George III, on 4 June 1738.

fourth position in ballet, a position either with the feet turned out and a step apart, the heels in line one behind the other (open or *ouvert* position), or with the feet turned out and a step apart, the toes of one foot in line with the heel of the other (crossed or *croisé* position).

Fourth Republic the republican government in France from 1945 to 1958.

Fourth Revolution the introduction of electronic and computerized instruction in schools in the 1970s (fourth by comparison with (1) the shift from home instruction to schoolteaching, (2) the adoption of the written word, (3) the invention of printing and the use of books; the term was coined in 1972 by Sir Eric Ashby in his report: 'The Fourth Revolution: Instructional Technology in Higher Education').

fourth wall in the theatre, the open front of the stage, regarded as representing a wall as solid as the other three to give an enclosed and naturalistic setting; if an actor addresses the audience he thus 'breaks' this wall.

Fourth World the poorest countries in the most undeveloped parts of the **Third World**.

QUOTATIONS AND SAYINGS

Four be the things I am wiser to know:
Idleness, sorrow, a friend, and a foe
(Dorothy Parker, *Inventory*, 1936)

Four be the things I'd been better without,
Love, curiosity, freckles, and doubt.
(Dorothy Parker, *Inventory*, 1936)

Four good mothers have four bad daughters: truth, hatred; prosperity, pride; security, peril; familiarity, contempt. (proverb)

Four legs good, two legs bad. (George Orwell, *Animal Farm*, 1945)

Four things greater than all things are, –
Women and Horses and Power and War.
(Rudyard Kipling, *Ballad of the King's Jest*, 1892)

Matthew, Mark, Luke, and John,
The bed be blest that I lie on.
Four angels to my bed,
Four angels round my head,
One to watch, and one to pray,
And two to bear my soul away.
(Thomas Ady, *A Candle in the Dark*, 1655)

The four pillars of government . . . (which are religion, justice, counsel, and treasure). (Francis Bacon, *Essays*, 1597)

There be three things which are too wonderful for me, yea four which I know not: The way of an eagle in the air; the way of a serpent upon a rock; the way of a ship in the midst of the sea; and the way of a man with a maid. (Bible: *Proverbs* 30:18–19)

We do not think the R's are only three.
We think Religion first of four to be.
(Tom Pollock, 19th century; compare the **three Rs**)

FIVE

EXPRESSIONS

BASED ON **FIVE**

Big Five (1) the UK, the USA, France, Italy and Japan, at the Paris Peace Conference of 1919, after the First World War; (2) China, France, the UK, the USSR and the USA, the major powers in the Second World War; (3) the five main banks before 1968: National Provincial, Westminster, Midland, Lloyds and Barclays (compare **Big Four**); (4) the heads of the five main departments of Scotland Yard: the Deputy Commissioner and the Assistant Commissioners in charge of 'A' (administration), 'B' (traffic and transport), 'C' (CID) and 'D' (recruitment and training) departments (superseded in 1980s when the structure was radically reorganized).

bunch of fives a fist, especially used for fighting (from its five fingers).

Cinque Ports the Kent and Sussex ports of Hastings, Sandwich, Dover, Romney and Hythe, granted special privileges from the 12th century for providing ships and men to defend the English Channel (other ports were added later).

Famous Five (1) the five main characters in the children's stories by Enid Blyton: two boys, Julian and Dick, two girls, Anne and 'George' (Georgina), all cousins, together with Timmy the Dog; (2) the five presenters who launched TV-am in 1983: David Frost, Michael Parkinson, Angela Rippon, Peter Jay and Anna Ford.

Five, The the five composers who united with the aim of creating a truly Russian school of music in *c.* 1875: Cui, Borodin, Balakirev, Mussorgsky and Rimsky-Korsakov (also known as the Mighty Five, but in Russian as the *Moguchaya kuchka*, 'mighty handful').

five agents in Chinese philosophy, the primary elements – water, fire, wood, metal and earth.

five-and-ten (five-and-dime) in US, a store where all articles were originally priced at 5 or 10 cents.

five-a-side a version of football with five players per side (compare **seven-a-side**).

five-barred gate a gate commonly found at the entrance to a field, with four horizontal wooden or metal bars and one transverse.

Five Blessings in Chinese art, the five bats that represent the blessings of longevity, wealth, serenity, virtue and an easy death.

Five Boroughs the Danish confederation of the towns (boroughs) of Derby, Leicester, Lincoln, Nottingham and Stamford in the 9th and 10th centuries.

five-by-five in aerospace jargon, excellent radio reception, based on a 1-to-5 scale for volume and audibility.

Five Civilized Nations (Five Civilized Tribes) the Indian tribes Cherokee, Choctaw, Chickasaw, Creek and Seminole, who were literate, had a representative government, practised trades other than farming, and who were forcibly resettled in the Indian Territory (modern Oklahoma) in the 1830s from their native southeast (the plains bordering the Gulf of Mexico).

five-day week the standard working week, Monday to Friday.

Five Fifths of Ireland the major federations established in the earliest historic times in Ireland: Ulaid (Ulster), Mide (Meath), Laigin (Leinster), Mumha (Munster) and Connachta (Connacht) (Irish, *Cuíg cuigí*) (compare **Six Counties**).

five-finger exercises musical scales practised on the piano by learners, hence generally something easy.

Five Freedoms of the Air the basic international agreement for air transport, with rights to (1) fly over a country without landing, (2) land for fuelling or repairs, (3) discharge passengers, mail and cargo from a home country in a foreign country, (4) pick up the same, (5) transport the same from one foreign country to another.

five-gaited of a show horse in US, trained to perform rack (halfway between trot and pace) and slow-gait as well as walk, trot and canter.

five Ks the five items traditionally worn or carried by Sikhs, each being symbolic: kuccha, kangha, kara, kesh and kirpan (Punjabi, *panch kakke*, 'five Ks').

Five Members the five members of the Long Parliament whom Charles I attempted to arrest in 1642: Pym, Hampden, Haselrig, Holles and Strode.

Five-Mile Act the act (of 1665, but repealed in 1812) prohibiting Nonconformist clergy from coming within five miles of any town or place where they had formerly ministered.

Five Nations (1) the 16th-century confederacy of North American Indian peoples living mainly in and around what is now New York State: Cayugas, Mohawks, Oneidas, Onondagas and Senecas; (2) Rudyard Kipling's term for the British Empire, comprising the Old Country, Canada, Australia, South Africa and India (see also **Five Nations Championship**).

Five Nations Championship in rugby union, the annual contest between the teams of England, Scotland, Ireland, Wales and France.

five-note scale in music, the pentatonic scale, representing the 1st, 2nd, 3rd, 5th and 6th degrees of the major diatonic scale, otherwise C, D, E, G, A, found in the music of many countries of the world, from Scotland to China and Japan.

five o'clock shadow the growth of beard or stubble that becomes visible on a man's face in the late afternoon.

Five Pillars of Islam the five essential obligations of the Moslem believer: (1) belief in one God, (2) prayer five times a day facing Mecca, (3) almsgiving, (4) observance of the Ramadan fast, (5) the hadj (pilgrimage to Mecca at least once in a lifetime).

fivepins in Canada (mainly), a bowling game using five pins.

Five Points of Calvin the five basic religious doctrines of the reformer John Calvin, enunciated in 1536: (1) the transcendence of God, (2) the total depravity of natural man, (3) the predestination of particular election (God chose some men for salvation before the world began), (4) the sole authority of the scriptures and the Holy Spirit, (5) the community must enforce the discipline of the church.

fiver in UK, a £5 note; in US, a $5 bill (see **five-spot**).

fives a ball game similar to squash but played with either bats or the hands (perhaps originally with teams of five players each).

five senses hearing, sight, smell, taste and touch (compare **sixth sense**, **seventh sense**).

five-spot in US, a $5 bill (see **fiver**).

five-star of the best quality, as applied to a hotel, restaurant, etc.

Five Towns in the works of Arnold Bennett, the Potteries towns (actually six in number): Bursley (really Burslem), Hanbridge (Hanley), Longshaw (Longton), Knype (Stoke-on-Trent) and Turnhill (Tunstall), all of which are now part of the city of Stoke-on-Trent (the sixth being Fenton).

five wits common sense, imagination, fantasy, estimation and memory, otherwise the faculties of the mind.

five Ws the five traditional questions a reporter should ask, ideally including the answers in his lead paragraph: Who? What? Why? When? Where?

five-year plan a planned period of economic development in Socialist countries, especially the USSR, where the first five-year plan began in 1929 (and the twelfth in 1986).

MI5 the former official and still popular name of the UK counter-intelligence agency (in full, Military Intelligence, Department 5; compare **MI6**).

take five to take a break (originally of five minutes).

Vancouver Five Julie Belmas, Gerry Hannah, Ann Hansen, Doug Stewart and Brent Taylor, young political activists convicted in 1982 for a series of bombings across Canada.

A LANGUAGE OF NUMBERS

Among the many artificial world languages, such as Esperanto, Volapük and Ido, there have been attempts to create a language solely from numbers. Probably the best known is the Gibson Code, named after the American Coast Artillery officer who devised it. In this, nouns start with 1, 2 or 3, verbs with 4, adjectives with 5, adverbs with 6, pronouns with 7, conjunctions with 8, and prepositions with 9. Tenses are indicated by adding 10, 20 or 30 to verbs, while nouns ending in an even number are indicated as plural, with odd numbers used to end singular nouns. In Gibson Code, 'The boy eats the red apple' works out as '5-111-409-10-5-516-2013', with '5' meaning 'the' and the central '10' denoting the present tense of the verb 'eat' ('409'). Gibson Code is thus similar to a number of commercial signal or cable codes, and was designed to be written, not spoken!

BASED ON FIFTH

Fifth Amendment of the US Constitution, stating that no person may be obliged to testify against himself and that no person may be tried for a second time on a charge for which he has already been acquitted.

Fifth Avenue the New York street associated with wealth, in both its stores and its luxury apartments.

fifth column a subversive group, originally one of Falangist supporters in Madrid in the Spanish Civil War (1936) who were prepared to join four columns of insurgents marching on the city.

Fifth Day the (former) Quaker name for Thursday.

fifth estate a grouping distinct from the **first**, **second**, **third** and **fourth estates**, variously defined as the trade unions, the BBC, or a 'scholarly guild' of those who, worldwide, have received a formal education at school and university.

fifth form (fifth year) schoolchildren aged 15 or 16 who are in their last (and examination) year at school (compare **sixth form**).

fifth-generation of computers, capable of performing several functions simultaneously (in parallel), introduced in the mid-1980s.

fifth leg in equestrianism, the ability of a horse to recover well after stumbling.

Fifth Monarchy Men a Puritan sect in 17th-century England, prominent in the Commonwealth and the Protectorate; they felt that the reign of Christ was at hand, this being the 5th monarchy after the Assyrian, Persian, Greek and Roman monarchies.

Fifth of November another name for Guy Fawkes Night or Bonfire Night, with fireworks and a bonfire, originally commemorating the 'Gunpowder Plot' of 5 November 1605, when Guy Fawkes and others planned to blow up the Houses of Parliament and James I on the day of the Opening of Parliament.

fifth position in ballet, a position with the legs crossed and the feet pointing in opposite directions, left and right, with the big toe of the back foot touching the heel of the one in front.

Fifth Republic the republican government of France since 1958.

fifth wheel a spare wheel for a four-wheeled vehicle, also any thing or person regarded as superfluous or inessential.

THE SEVEN PILLARS OF WISDOM

T E Lawrence ('Lawrence of Arabia') is best remembered for his book *The Seven Pillars of Wisdom*, describing the rebellion of the Arabs against the Turks in the First World War, and his own part in it. But what is the significance of the title? Lawrence adapted it from the biblical line, 'Wisdom hath builded her house, she hath hewn out her seven pillars' (*Proverbs* 9:1). He originally used it for an earlier work, a travel book about seven cities: Cairo, Smyrna, Constantinople, Beirut, Aleppo, Damascus and Medina, with the Crusades as its underlying theme. But he was dissatisfied with the book and burnt the manuscript. He kept the title, however, for the later work, and it can be regarded as apt to describe the 'pillars' of physical, mental and spiritual strength that he needed to sustain him in his desert campaign, with the eventual 'wisdom' his cumulative experience derived during the course of it.

QUOTATIONS AND SAYINGS

And five of them were foolish, and five were wise. (Parable of the Ten Virgins, Bible: *Matthew* 25:2)

Five minutes! Zounds! I have been five minutes too late all my life-time! (opening line of Mrs Hannah Cowley, *The Belle's Stratagem*, 1780)

Full fathom five thy father lies. (Shakespeare, *The Tempest*, 1611)

He that would thrive
Must rise at five;
He that hath thriven
May lie till seven.
(John Clarke, *Paraemiologia Anglo-Latina*, 1639)

If all be true that I do think,
There are five reasons we should drink:
Good wine – a friend – or being dry –
Or lest we should be by and by –
Or any other reason why.
(Henry Aldrich, *Five Reasons for Drinking*, 17th century)

If your lips would keep from slips,
Five things observe with care:
To whom you speak, of whom you speak;
And how, and when, and where.
(William Edward Norris, quoted in his *Thirlby Hall*, 1883)

Oranges and lemons, say the bells of St Clement's,
You owe me five farthings, say the bells of St Martin's.
(nursery rhyme and children's action game)

We have heard it said that five per cent is the natural interest of money. (Macaulay, *Literary Essays*, 1843)

We have here but five loaves, and two fishes. (Feeding of the Five Thousand, Bible: *Matthew* 14:17)

SIX

EXPRESSIONS

BASED ON SIX

at sixes and sevens in disorder (the phrase is either from dicing, with dice having a higher value than today, or else a corruption of *to set on cinque and sice*, that is, 'to gamble on five and six', regarded as the most risky bets).

Birmingham Six the six terrorists who planted bombs in Birmingham in 1974, and who were sentenced to life imprisonment the following year: Patrick Joseph Hill, Robert Gerald Hunter, Noel Richard McIlkenny, William Power, John Walker, Hugh Daniel Callaghan.

deep six in US, a euphemism for death, the grave (perhaps a reference to burial at sea six fathoms deep) (compare **six feet under**).

go (be hit) for six to be quite defeated (from the powerful cricket hit that scores 6: see **six**).

hit on all six to be on excellent form (like the engine of a large car, in which all six cylinders are working perfectly; compare **fire on all four cylinders**).

Les Six the French composers who banded together in 1918 under Cocteau and Satie as a reaction to the romantic school and to promote modern music: Honegger, Milhaud, Poulenc, Durey, Auric, Tailleferre.

MI6 the former official and still popular name for the British Secret Intelligence Service (the SIS), the equivalent of the US CIA (in full, Military Intelligence, Department 6; compare **MI5**).

San Quentin Six the six men killed in an attempted escape from California State Prison, San Quentin, in 1971: George Jackson, John Lyne, Ronald Kane, Frank DeLeon, Gere Graham, Paul Krasnes.

Sharpeville Six the six black persons sentenced to death for alleged complicity in the mob murder of a black township official in Sharpeville, South Africa, in 1984; they were reprieved in 1988: Duma Joshua Khumalo, Francis Don Mokgesi, Reginald Mojalefa Sefatsa, Reid Malebo Mokoena, Oupa Moses Diniso, Theresa Ramashamola.

six (1) a score of six runs in cricket, gained by hitting the ball over the boundary without its touching the ground; (2) a division of a Brownie Guide or Cub Scout pack (containing six).

Six, The the six countries who were the original partners in the European Coal and Steel Community (from 1951), the European Economic Community (from 1957) and Euratom (from 1957): Belgium, France, West Germany, Italy, The Netherlands, Luxembourg (compare the **Seven** (2), the **Ten** (1), the **Twelve** (2)).

Six Articles the statute passed in 1539 to secure conformity in matters of religion, comprising: (1) transubstantiation, (2) the sufficiency of communion in one kind, (3) clerical celibacy, (4) the obligation imposed by monastic vows, (5) the propriety of private masses, (6) the necessity of auricular confession; the statute, nicknamed 'the whip with six strings', was repealed in 1547.

Six Counties the counties of Northern Ireland: Antrim, Armagh, Down, Fermanagh, Londonderry (Derry), Tyrone.

six-day race an indoor cycle race lasting six days, popular in the US, where first introduced in 1891 at Madison Square Garden, New York.

Six Day War the Arab-Israeli war in the Middle East from 5 to 10 June 1967, in which Israel defeated Egypt, Jordan and Syria and occupied much Arab territory.

six-eight time in music, compound duple time, with six quaver (eighth note) beats to a bar, and indicated as $\frac{6}{8}$.

sixer the leader of a **six** (2).

six feet under dead and buried (compare **deep six**).

six-footer a tall person, 6ft (1.8m) or more in height.

six-foot way the strip of ground between two railway tracks (compare **four-foot way**).

six-man football in US, a variety of American football developed in 1934 for use in small **secondary schools** which could not produce a full 11-man team.

Six Nations the Indian confederacy of the Cayugas, Mohawks, Oneidas, Onondagas, Senecas and Tuscaroras, when this last race joined the **Five Nations** (1) in 1715.

six o'clock swill in Australia and New Zealand, a period of heavy drinking (originally at a time when hotels had to close their bars at 6 p.m.).

six of one and half a dozen of the other equal, equally acceptable (of two alternatives).

six of the best a severe beating, usually with six strokes of a cane, as formerly in UK schools ('best' as powerful).

six-pack a pack of six items to be sold as a single unit, for example six cans of beer.

sixpence a small silver-coloured coin worth six pre-decimal pence, not minted after 1970, hence anything small.

six pips the BBC time signal, broadcast on the hour, and consisting of five short 'pips' (bleeps) and one long one, the latter on the exact hour.

Six Points of Ritualism the religious rites proposed by the Anglo-Catholic Oxford Movement in the 1870s, comprising: (1) altar lights, (2) eucharistic vestments, (3) a mixed chalice, (4) incense, (5) unleavened bread, (6) the eastward position for the celebrant.

Six-Principle Baptists an Arminian sect founded in about 1639 who based their creed on the six principles enunciated in *Hebrews* 6:1–2 – repentance, faith, baptism, the laying on of hands, the resurrection of the dead, eternal judgment.

six-shooter (six-gun) in US, a revolver with six chambers.

six to four on likely to occur (betting jargon).

six-yard line in football, the line that marks the limit of the goal area.

BASED ON **SIXTH**

Sixth Avenue the Avenue of the Americas, New York, famous for its glass- and steel-structured buildings.

Sixth Commandment 'Thou shalt do no murder', one of the key **Ten Commandments**.

Sixth Day the (former) Quaker name for Friday.

sixth form the most senior class in a **secondary school**, usually above the legal school-leaving age (16), in which students work for A- or AS-levels (for admittance to university or higher education).

sixth-form college a college offering education at **sixth-form** level, especially for students from a school without this class.

sixth-generation of computers, using artificial intelligence in the form of biological organisms and tissue to perform basic logical functions, introduced from the late 1980s.

sixth sense a supposed sense of intuition or clairvoyance, additional to the **five senses** (compare **seventh sense**).

QUOTATIONS AND SAYINGS

Above it stood the seraphims; each one had six wings; with twain he covered his face, and with twain he covered his feet, and with twain he did fly (Bible: *Isaiah* 6:1)

A youth who likes to study will in the end succeed. To begin with he should know that there are Six Essentials in painting. The first is called *spirit*; the second, *rhythm*; the third, *thought*; the fourth, *scenery*; the fifth, the *brush*; and the last is the *ink*. (Ching Hao, *Notes on Brushwork*, 10th century)

But now I'm six, I'm clever as clever,
So I think I'll be six for ever and ever.
(A A Milne, *Now We Are Six*, 1927)

I keep six honest serving men
(They taught me all I knew);
Their names are What and Why and When
And How and Where and Who.
(Rudyard Kipling, *The Just-So Stories*, 1902)

In six days the Lord made heaven and earth, the sea, and all that in them is. (Book of Common Prayer, 4th Commandment)

Sing a song of sixpence. (nursery rhyme)

Six days shalt thou labour, and do all that thou hast to do. (Book of Common Prayer, 4th Commandment)

Six hours in sleep, in law's grave study six,
Four spend in prayer, the rest on Nature fix.
(Sir Edward Coke, translation of *The Pandects*, 6th century)

Why, sometimes I've believed as many as six impossible things before breakfast. (Lewis Carroll, *Through the Looking-Glass*, 1872)

SEVEN

EXPRESSIONS

BASED ON SEVEN

Chapter 7 in US, a provision of the US Federal Bankruptcy Act for the relief of insolvent debtors and their creditors (compare **Chapter 11**, **Chapter 13**).

Chicago Seven the seven persons convicted in 1970 of conspiracy to incite a riot and of causing other disturbances at the Democratic National Convention, Chicago, in 1968: Rennie Davie, David T. Dellinger, John Forines, Thomas Hayden, Abbie Hoffman, Jerry C. Rubin, Lee Weiner (originally called the Chicago Eight, but the eighth defendant, Bobby Seale, was tried separately).

Dance of the Seven Veils the dance traditionally performed by Salome, especially in Oscar Wilde's play *Salome* (1894), familiar from Richard Strauss's opera of the same name (1905) based on it; Salome gradually removes the seven veils she is wearing.

Group of Seven the leading industrial nations of the world, whose finance ministers meet regularly to discuss international monetary and economic strategies: UK, West Germany, USA, Japan, France, Italy, Canada.

L7 in US, former Hollywood slang (in the 1950s) for a 'square' or traditionally-minded person (from the square formed by the forefinger and thumb, the left hand giving 'L' and the right '7'.

Seven, the (1) the seven 'men of honest report' chosen by the apostles to be the first deacons (*Acts* 6:5): Stephen, Philip, Prochorus, Nicanor, Timon, Parmenas, Nicolas; (2) the original members of the European Free Trade Association (EFTA) in 1959: Denmark, Norway, Sweden, Austria, Portugal, Switzerland, UK (compare the **Six**, the **Ten**).

Seven against Thebes in Greek mythology, the seven members of the expedition to regain for Polynices his share in the throne of Thebes from his usurping brother Eteocles: Polynices, Adrastus, Amphiaraus, Capaneus, Hippomedon, Tydeus, Parthenopaeus.

seven-a-side another name for **sevens** (see below).

Seven Bishops the seven bishops who petitioned James II in 1688 against the order to have his second Declaration of Indulgence read in every church on two successive Sundays: Archbishop Sancroft of Canterbury, Bishops Lloyd of St Asaph, Turner of Ely, Ken of Bath and Wells, White of Peterborough, Lake of Chichester, Trelawney of Bristol.

Seven Brothers the seven sons of St Felicity, who with their mother were arrested by Antoninus Pius, for refusing to sacrifice to the Roman gods, and executed: Felix, Philip, Martial, Vitalis, Alexander, Silvanus, Januarius (the story may well be apocryphal).

Seven Champions of Christendom the patron saints of England (George), Scotland (Andrew), Wales (David), Ireland (Patrick), France (Denys), Spain (James), Italy (Anthony).

Seven Churches of Asia those listed in *Revelation* 1:11: Ephesus, Smyrna, Pergamos, Thyatira, Sardis, Philadelphia, Laodicea.

Seven Days' Battles the series of engagements from 25 June to 1 July 1862, in which the Confederate Army under General Robert E. Lee prevented the Union troops under General George B. McClennan from capturing the Confederate capital at Richmond, Virginia, USA.

Seven Founders the seven men who in 1233 established themselves as hermits on Monte Senario, near Florence, Italy, so founding an order of mendicant friars, the Servants of Mary (Servites): St Buonfiglio Monaldo, St Alexis Falconieri, St John Bonaiuncta, St Manettus dell' Antella, St Amadeus degli Amidei, St Hugh Ugucionne, St Sosthenes Sostegno (also known as the Seven Servite Founders).

Seven Gifts of the Holy Spirit in the Christian religion: wisdom, understanding, counsel, fortitude, knowledge, righteousness, fear of the Lord (but in *I Corinthians* 12:4–11 the gifts are given as: wisdom, knowledge, faith, healing, the working of miracles, prophecy, the discerning of spirits, the speaking of languages, the interpreting of languages – nine in number).

Seven Hills of Rome the hills on which ancient Rome was built: Palatine, Capitoline, Quirinal, Caelian, Aventine, Esquiline, Viminal.

Seven Joys of Mary in the Christian religion, seven joyful events in the life of the Virgin Mary: the Annunciation, the Visitation, the Nativity, the

Epiphany, the Finding in the Temple, the Resurrection, the Ascension (compare the **Seven Sorrows of Mary**).

seven-league boots the boots worn by a giant in folk tales, enabling him to take strides of seven leagues (21 miles).

Seven Liberal Arts in Ancient Rome, the combined Quadrivium (arithmetic, music, geography, astronomy) and Trivium (grammar, rhetoric, logic), also jointly known as the Seven Sciences.

Seven Names of God El, Elohim, Adonai, Yahweh, Ehyeh-Asher-Ehyeh, Shaddai, Zaba'ot.

seven orifices the seven natural openings of the human body: 2 ears, 2 nostrils, mouth, penis or vagina, anus (compare **eighth orifice**).

Seven Planets according to ancient beliefs: the Sun, the Moon, Mars, Mercury, Jupiter, Venus, Saturn (which gave the names of the days of the week in some languages, see p. 34).

sevens a series of rugby matches played with seven players a side instead of the usual 15 (also called **seven-a-side**).

Seven Sacraments in the Christian religion, the sacraments of baptism, confirmation, the eucharist, penance, extreme unction, orders, matrimony (listed in the 25th of the **Thirty-Nine Articles** in the Book of Common Prayer).

Seven Sages (of Greece) Solon of Athens, Thales of Miletus, Pittacus of Mitylene, Bias of Priene in Caria, Chilon of Sparta, Cleobulus tyrant of Lindus in Rhodes, Periander tyrant of Corinth.

seven seas the seven oceans of the world: North Pacific, South Pacific, North Atlantic, South Atlantic, Arctic, Antarctic, Indian.

Seven Senses according to *Ecclesiasticus* 17:5—animation, feeling, speech, taste, sight, hearing, smelling (compare **Five Senses**).

Seven Sisters (1) the Pleiades, in Greek mythology, the seven daughters of Atlas and Pleione: Alcyone, Merope, Maia, Electra, Taygete, Sterope, Celaeno, preserved as stars in the sky; (2) the seven international oil companies (as at 1979): Exxon, Mobil, Gulf, Standard Oil of California, Texaco, BP, Royal Dutch Shell; (3) the line of chalk cliffs in Sussex, east of Beachy Head: Went Hill Brow, Baily's Brow, Flagstaff Point, Brass Point, Rough Brow, Short Brow, Haven Brow; (4) in US, the seven elite women's colleges: Bernard, Bryn Mawr, Mount Holyoake, Radcliffe, Smith, Vassar, Wellesley; (see also the place names Seven Sisters, p. 150).

Seven Sleepers (of Ephesus) the seven Christian boys walled up in a cave by the Emperor Decius in AD 250, when according to legend they slept for 187 years (the number varies): Constantius, Dionysius, Joannes, Maximianus, Malchus, Martinianus, Serapion (other accounts give other names).

Seven Sorrows of Mary in the Christian religion, seven sorrowful events in the life of the Virgin Mary: Simeon's prophecy, the Flight into Egypt, the loss of the Holy Child, the road to Calvary, the Crucifixion, the Descent (Taking Down) from the Cross, the Entombment (compare **Seven Joys of Mary**).

seven stages of drunkenness verbose, grandiose, amicose, bellicose, morose, stuporous, comatose.

seven-up a card game in which the card turned up as the lead to each round determines the trump suit, with a total of seven points needed to win a game.

Seven Virtues faith, hope, charity, justice, fortitude, prudence, temperance.

Seven Weeks War the war of 14 June to 26 July 1866, between Austria and Prussia.

Seven Wonders of the World the seven most wondrous objects of the ancient world: the Pyramids of Egypt, the Hanging Gardens of

NUMERICAL SEQUELS

Some popular authors write sequential novels (based on the same central character or characters) that have numerical titles running in sequence. In his 'Colonel Granby' books, for example, Francis Beeding (pseudonym of John L. Palmer and Hilary St George Saunders) wrote *One Sane Man, The Two Undertakers, Three Fishers, Four Armourers, Five Flamboys, Six Proud Walkers, Seven Sleepers, Eight Crooked Trenches, Nine Waxed Faces, The Ten Holy Horrors, Eleven Were Brave, The Twelve Disguises* and *There Are Thirteen*. In the world of children's books, Enid Blyton not only wrote about the 'Adventurous Four', the 'Famous Five', the 'Six Cousins' and the 'Secret Seven', but had sequentially numbered titles in a single series, such as the 'Malory Towers' books, following a group of children as they progress through school: *First Term at Malory Towers, Second Form at Malory Towers, Third Year at Malory Towers, Upper Fourth at Malory Towers, In the Fifth at Malory Towers* and, finally, *Last Term at Malory Towers*. No doubt she felt that sixth formers were too senior to feature in school stories!

Babylon, Phidias' statue of Zeus at Olympia, the Temple of Artemis at Ephesus, the Mausoleum of Halicarnassus, the Colossus of Rhodes, the Pharos (lighthouse) of Alexandria; a later list gives the Wonders as: the Coliseum of Rome, the Catacombs of Alexandria, the Great Wall of China, Stonehenge, the Leaning Tower of Pisa, the Porcelain Tower of Nanking, the Mosque of San Sophia at Constantinople (Istanbul).

Seven Wonders of Wales the mountains of Snowdon, the churchyard of Overton, the bells of Gresford church, the bridge at Llangollen, Wrexham steeple (actually a tower), the waterfall at Pistyll Rhaeadr, St Winefride's well at Holywell.

Seven Wonders of Wiltshire Stonehenge, Avebury, Silbury Hill, Old Sarum, Salisbury Cathedral, the Westbury White Horse, Wilton House.

Seven Words from the Cross the last sayings of Jesus: 'Father, forgive them; for they know not what they do' (*Luke* 23:34), 'Today shalt thou be with me in paradise' (*Luke* 23:43), 'Woman, behold thy son!' (*John* 19:26), 'My God, my God, why hast thou forsaken me?' (*Matthew* 27:46), 'I thirst' (*John* 19:28), 'It is finished' (*John* 19:30), 'Father, into thy hands I commend my spirit' (*Luke* 23:46).

Seven Works of Mercy in the Christian religion, the seven corporal works (tend the sick, feed the hungry, give drink to the thirsty, clothe the naked, harbour the stranger, minister to prisoners, bury the dead) (*Matthew* 25:35–45) or the seven spiritual works (convert the sinner, instruct the ignorant, counsel the doubtful, comfort the sorrowing, bear wrongs patiently, forgive injuries, pray for the living and the dead).

seven-yard line in hockey, the line running parallel to the sideline, seven yards into the field; players must be behind the line to receive a roll-in.

seven-year itch a tendency to infidelity said to begin after seven years of married life.

Seven Years War the war of 1756–63 between Britain and Prussia against France and Austria.

Siberian Seven the Soviet Pentecostalists who took refuge in the US Embassy, Moscow in 1978: Timofey and Maria Chmykhalova, and Avgustina, Lidia, Lilia, Lyuba and Pyotr Vashchenko; the last of the seven was released in 1983.

Watergate Seven the seven men convicted of breaking into and 'bugging' the Democratic National Committee Headquarters in the Watergate office building, Washington, DC, in 1972: Bernard L. Baker, Virgilio R. Gonzalez, E. Howard Hunt, Jr, G. Gordon Liddy, James W. McCord, Eugenio Martinez, Frank A. Sturgis.

BASED ON **SEVENTH**

Seventh Avenue the New York street that is the centre of the clothing, tailoring and dressmaking industry.

Seventh Commandment the injunction not to commit adultery, as one of the **Ten Commandments**.

Seventh Day the (former) Quaker name for Saturday.

Seventh-Day Adventists a religious sect of US origin, founded in 1831, and observing Saturday (see **Seventh Day**) as their Sabbath.

seventh heaven the height of perfection or bliss (in Islam, where the seventh heaven is formed of divine light and is ruled by Abraham).

seventh sense a sense of humour (in addition to the **five senses** and a **sixth sense**).

QUOTATIONS AND SAYINGS

And Jacob served seven years for Rachel; and they seemed unto him but a few days, for the love he had to her (Bible: *Genesis* 29:20)

As I was going to St Ives,
I met a man with seven wives,
Each wife had seven sacks,
Each sack had seven cats,
Each cat had seven kits:
Kits, cats, sacks, and wives,
How many were there going to St Ives?
(nursery rhyme: the answer is, of course, 1!)

Give me a child for the first seven years and you may do what you like with him afterwards. (Jesuit maxim)

If you keep a thing for seven years, you are sure to find a use for it. (Walter Scott, *Woodstock*, 1826)

Lone and alone she lies,
Poor Miss 7,
Five steep flights from the earth,
And one from heaven.
(Walter de la Mare, *Poor 'Miss 7'*, 1913)

One man in his time plays many parts,
His acts being seven ages.
(Shakespeare, *As You Like It*, 1599; the seven ages of man are: infant, schoolboy, lover, soldier, justice, pantaloon, second childishness)

Seven is a good handy figure in its way, picturesque, with a savour of the mythical; one might say that it is more filling to the spirit than a dull academic half-dozen. (Thomas Mann, *The Magic Mountain*, 1924)

There's no dew left on the daisies and clover,
There's no rain left in heaven;
I've said my 'seven times' over and over,
Seven times one are seven.
(Jean Ingelow, *Songs of Seven: Seven Times One*, 19th century)

The year's at the spring,
And day's at the morn;
Morning's at seven,
The hill-side's dew-pearled.
(Robert Browning, *Pippa Passes*, 1841)

Those who divide life into periods of seven years are not far wrong, and we ought to keep to the divisions that nature makes. (Aristotle, *Politics*, 4th century BC)

EIGHT

EXPRESSIONS

BASED ON EIGHT

Big Eight in US, the major firms of certified public accountants: Arthur Andersen & Co.; Arthur Young & Co.; Coopers & Lybrand; Deloitte, Haskins & Sells; Ernst & Whinney; Peat, Marwick, Mitchell & Co.; Price, Waterhouse; Touche, Ross & Co.

Eight, the the group of US painters also known as the 'Ashcan School', who exhibited together only once, in New York in 1908, but who established an important realist trend in painting: Robert Henri, Everett Shinn, John Sloan, Arthur BDavies, Ernest Lawson, Maurice Prendergast, George Luks, William J. Glackens.

eightball (1) in US and Canada, the black ball in the game of pool, which is marked with a No. 8; (2) an inefficient or maladjusted person (perhaps from sense (1), since a player hitting the eightball prematurely with the cue ball loses points, and to be behind the eightball is to be in a difficult position, but with the phrase also possibly influenced by **Section 8**, see below).

eight-day clock a clock that will go for eight days without winding.

Eightfold Path in Buddhism, the summary, together with the **Four Noble Truths**, of the whole of Buddhist teaching: (1) right understanding, (2) right thought, (3) right speech, (4) right action, (5) right livelihood, (6) right effort, (7) right mindfulness, (8) right concentration.

eight-hour day the standard '9-to-5' working day of 8 hours.

Eight Points the points of the Atlantic Charter of 1941, issued jointly by the UK and USA, stating that (1) neither nation sought aggrandizement, (2) there must be no territorial changes without the agreement of the peoples concerned, (3) there must be respect for every people's right to choose its own form of government, (4) there must be equal access for all states to trade and raw materials, (5) there must be worldwide collaboration to improve labour standards, economic progress and social security, (6) there must be peace for all nations after the destruction of 'Nazi tyranny', (7) there must be freedom of the seas under such peace, and (8) potential aggressors must disarm.

Eights the annual rowing ('bumping') races on the river between crews of eight rowers from each college at Oxford; the races take place in June in 'Eights Week'.

POET AND PEDANT

The 19th-century mathematician Charles Babbage, who invented a complex calculating machine that was never constructed for lack of financial backing, took the poet Tennyson to task for the latter's lines:

Every minute dies a man,
Every minute one is born.

Babbage pointed out that the world's population was in fact constantly increasing: 'I would therefore take the liberty of suggesting that in the next edition of your excellent poem the erroneous calculation to which I refer should be corrected as follows:

Every moment dies a man
And one and a sixteenth is born.

He added that this figure was actually a concession to metre, as the true ratio was 1:167. Tennyson did not adopt his numerical suggestion, although he *did* change 'minute' to 'moment'.

eightsome a group of eight people, especially a Scottish reel danced by eight dancers.

figure of eight an outline of the figure '8' traced or followed by someone or something, such as a skater or an aeroplane.

New York Eight a group of eight activists in Brooklyn, New York, charged in 1984 with conspiring to rob armoured cars and other offences: Ruth Carter, Coltrain Chimurenga, Omowale Clay, Yvette Kelly, Viola Plummer, Roger Taylor, Roger Wareham, Howard Bonds; a ninth activist, Pepe Ríos, was later indicted on the same charges, but as posters and T-shirts had been printed with 'New York 8', it was decided to add '+' for Ríos, instead of altering '8' to '9'.

No. 8s working dress in the Royal Navy.

number 8 in rugby union, the player in the back row of the scrum (as the 8th after the two prop forwards, hooker, two lock forwards and two wing forwards).

one over the eight drunk (that is, having more than eight glasses of beer).

piece of eight a former Spanish coin ('dollar') worth 8 reals.

Section 8 a US Army regulation in the Second World War that provided for the discharge of a person found physically or mentally unfit to serve (see also **eightball** and compare **ward 8**).

ward 8 in US military slang, the hospital ward where the insane are treated (compare **Section 8**).

BASED ON **EIGHTH**

Eighth Army the grouping of British forces formed in the Western Desert in 1941, separating the British forces there from those in Egypt, and fighting in North Africa under various commanders (eventually Montgomery) to join with US forces in the invasion of Sicily and the Italian campaign (the number was arbitrary, as a deliberate policy to confuse the enemy, and there was no 'Seventh' or earlier army).

eighth note in music, an alternative name (mostly in US and Canada) for a quaver (as one-eighth the length of a whole note or semibreve).

eighth orifice a term sometimes used by crime and detective fiction writers for a wound to the body, especially a bullet or knife wound (compare **seven orifices**).

eighth wonder of the world something wonderful, unique or unexpected (often used sarcastically) (compare **Seven Wonders of the World**).

QUOTATIONS AND SAYINGS

Look, here's the warrant, Claudio, for thy death:
'Tis now dead midnight, and by eight tomorrow
Thou must be made immortal.

(Shakespeare, *Measure for Measure*, 1604; 8 a.m. was long the traditional time for hanging a criminal)

We want eight, and we won't wait. (public demand for eight Dreadnoughts to be built, quoted by George Wyndham, 1909)

When eight days were accomplished for the circumcising of the child, his name was called JESUS. (Bible: *Luke* 2:21; in the Christian church, the Feast of the Circumcision, now often called the Naming of Jesus, falls on 1 January, the eighth day, or Octave, from 25 December, Christmas Day)

NINE

EXPRESSIONS

BASED ON **NINE**

cat-o'-nine-tails a whip with nine lashes (perhaps an oblique reference to a cat, which traditionally had **nine lives**).

Catonsville Nine the nine Vietnam War protesters charged with destroying government property and impeding Selective Service procedures in Catonsville, Maryland, in 1967: Daniel Berrigan, Philip E. Berrigan, David Durst, John Hogan, Thomas P. Lewis, Marjorie B. Melville, Thomas Melville, George Mische Mary Moylan.

dressed up to the nines elaborately or extravagantly dressed or overdressed (probably from the mystic use of nine to express perfection; compare **on cloud nine**).

go down like ninepins to succumb in large numbers, as in an epidemic (or like **ninepins**).

K-9 Corps in US, the dog defence corps in the Second World War (punning version of 'canine', and a stock name for a robot dog elsewhere).

look nine ways to squint.

nice as ninepence very nice (and similar

phrases such as 'nimble as ninepence', 'right as ninepence') (perhaps influenced by **ninepins**).

Nine, the the nine Muses in Greek mythology: Calliope (the Muse of epic poetry), Clio (heroic poetry), Erato (love poetry), Euterpe (music), Melpomene (tragedy), Polyhymnia (sacred poetry), Terpsichore (choral song and dance), Thalia (comedy), Urania (astonomy).

Nine Days Queen Lady Jane Grey, who was queen for the 9 days 10 to 19 July 1553, when she was imprisoned on a charge of high treason and later executed, aged 16.

nine days wonder something that arouses great but shortlived interest (said to originate from the time, 9 days, taken by puppies and kittens to open their eyes after birth).

Nine First Fridays in the Roman Catholic church, the first Friday in each of nine consecutive months, marked by receiving the eucharist (based on the promise which Christ is said to have made to St Margaret Mary Alacoque, the 16th-century founder of devotion to the Sacred Heart of Jesus, that unusual graces would be given to all who received it thus).

Nine Lessons and Carols (Festival of Nine Lessons and Carols) the Christmas carol service held in many churches, chapels and cathedrals, with nine carols alternating with nine Bible readings (the practice originated in Truro Cathedral in 1880, but was popularized by King's College, Cambridge, which first held it in 1918 and first broadcast it in 1928).

nine lives the proverbial number of lives of a cat, which has an unusual ability to survive accidents and 'scrapes', partly because of its great agility.

nine men's morris an old game similar to draughts (checkers), played on a board or greensward; its two players have nine 'men' (pieces) each.

999 in UK, the telephone number of the emergency services (police, fire, ambulance, etc.) (said to have been selected as a number easy to find and dial in the dark or smoke).

nine o'clock news the traditional time for the BBC evening news broadcast on radio (from 1927 to 1960) and television (from 1970).

ninepence in the shilling mentally retarded, 'not all there' (only partly sane or lucid).

ninepins skittles (with nine pins to be knocked down).

nine points of the law often applied to possession, as being nine-tenths of the total, and having every advantage but that of actual ownership or legal right, but also defined separately as: (1) a good deal of money, (2) a good deal of patience, (3) a good cause, (4) a good lawyer, (5) a good counsel, (6) good witnesses, (7) a good jury, (8) a good judge, (9) good luck, these being the ideal nine requirements of a successful legal case or trial.

Nine Spheres in the Ptolemaic system of astronomy, with Earth as the centre of the universe, the nine spheres of invisible space, the first seven carrying the **Seven Planets**, the eighth the Starry Sphere, with the fixed stars, and the ninth, the Crystalline Sphere, added later to account for the precession of the equinoxes.

nine tailors a method of change-ringing, made famous by Dorothy L. Sayers' murder story, *The Nine Tailors*, which includes a study of the art (with 'tailor' really 'teller', a stroke on a bell at a funeral, traditionally six for a woman and nine (3 times 3) for a man).

nine times out of ten very often, usually, more often than not.

nine to five (9 to 5) the standard **eight-hour day** of a person at work.

Nine Worthies nine famous men in history, either (1) Hector, Alexander, Julius Caesar (three Gentiles), Joshua, David, Judas Maccabaeus (three Jews) and King Arthur, Charlemagne, Godfrey of Bouillon (three Christians), or (2) the nine privy councillors of the reign of William III: Devonshire, Dorset, Monmouth, Edward Russell (Whigs), and Carmarthen, Pembroke, Nottingham, Marlborough, Lowther (Tories) (compare **Nine Worthy Women**).

Nine Worthy Women nine famous women in history or mythology: Minerva, Semiramis, Tomyris, Jael, Deborah, Judith, Britomart, Elizabeth (Isabella) of Aragon, Johanna of Naples (compare **Nine Worthies**).

on cloud nine very happy, delighted, elated (probably from the idea of nine as a perfect number, compare **dressed up to the nines**, **seventh heaven**).

rule of nines the method of measuring the extent of a burn on the human body as a percentage of the whole: head and arms are 9 per cent each, front of trunk and back of trunk and

legs are 18 per cent; these are then added together, giving a total of 99 per cent, to which is added 1 per cent for the perineum.

QUOTATIONS AND SAYINGS

And her shoes were number nine.
(Clementine, the miner's daughter, in the song by Percy Montrose, 19th century; see also **Forty-Niner**)

and then there were nine.
(Indians, niggers, etc., in nursery rhyme 'countdown')

A stitch in time saves nine. (proverb)

A woman hath nine lives like a cat. (John Heywood, *Proverbs*, 16th century, compare **nine lives**)

Lars Porsena of Clusium
By the nine gods he swore.
(Macaulay, *Lays of Ancient Rome*, 1842; the nine Etruscan gods were: Juno, Minerva, Tinia, Hercules, Mars, Saturn, Summanus, Vedius, Vulcan)

Nine bean-rows will I have there.
(W B Yeats, *The Lake Isle of Innisfree*, 1893)

Nine tailors make a man. (proverb)

Pease-pudding hot, pease-pudding cold,
Pease-pudding in the pot, nine days old.
(nursery rhyme)

Weary se'nnights nine times nine
Shall he dwindle, peak and pine.
(Witches' curse in Shakespeare's, *Macbeth*, 1606; 'se'nnights' are weeks)

Were there not ten cleansed? But where are the nine? (Bible: *Luke* 17:17)

TEN

EXPRESSIONS

BASED ON TEN

Club of Ten the sponsor of advertisements in European and US newspapers from 1973 to 1978 defending South African policies (see also **Group of Ten**).

Council of Ten the council of ten members instituted in Venice in the 14th century to watch over public order, and remaining a government organ until 1797.

count (up) to ten to pause before saying something that could be rash or hasty (some people actually do count to ten when under provocation).

give her ten an instruction to the crew of a rowing boat to row flat out for ten strokes.

Group of Ten (Club of Ten) the ten countries who after the UK balance of payments crisis in 1961 agreed to pay extra quotas to the International Monetary Fund in order to increase its lending resources: USA, UK, Canada, France, West Germany, Holland, Belgium, Italy, Sweden, Japan.

Number 10 see **10 Downing Street**.

Number 10 Club a nickname of the Institute of Directors, an organization of business leaders founded in 1903 (from its address to 1978, at No. 10 Belgrave Square, London; it is now in Pall Mall).

Ten, the (1) the ten countries that grouped to form the European Economic Community (Common Market) after 1981, when Greece joined the existing nine: Belgium, Denmark, France, West Germany, Greece, Ireland, Italy, Luxembourg, Netherlands, UK: when Spain and Portugal also joined in 1986, the 'ten' became 12 (compare the **Six**, the **Seven**); (2) the group of ten US painters who first exhibited together in New York in 1898 with the aim of drawing attention to their work, which was not in itself radically distinctive: Childe Hassam, John Henry Twachtman, J. Alden Weir, Thomas

W. Dewing, Joseph De Camp, Frank W. Benson, Willard Leroy Metcalf, Edmund Charles Tarbell, Robert Reid, E E Simmons.

ten-acre block in New Zealand, a block of subdivided farming land, originally 10 acres in area, which is usually within commuting distance of a city and which provides a semi-rural way of life.

Ten-cent Jimmy a nickname of James Buchanan, 15th US President, for his advocacy of low tariffs and low wages.

ten-code the main CB radio code of signals, originally used by US police and all beginning with the number 10 as a prefix (see p. 79).

Ten Commandments the ten basic obligations of man towards God and his fellow men, as delivered to Moses on Mount Sinai engraved on two tablets of stone: (1) Thou shalt have no other gods before me, (2) Thou shalt not make . . . any graven image, (3) Thou shalt not take the name of the Lord thy God in vain, (4) Remember the sabbath day, to keep it holy, (5) Honour thy father and thy mother, (6) Thou shalt not kill, (7) Thou shalt not commit adultery, (8) Thou shalt not steal, (9) Thou shalt not bear false witness against thy neighbour, (10) Thou shalt not covet (*Exodus* 20:1–17 gives full wording) (compare **eleventh commandment**).

Ten Days of Penitence the Jewish feast of Yamim Noraim, the first ten days of the religious year, comprising the three High Holy Days of Rosh Hashana and Yom Kippur and the days in between (Hebrew name is *aseret yeme teshuva*, 'ten days of penitence').

'Ten Days That Shook the World' a term for the Russian Revolution of 1917 (title of book describing it by John Reed, 1919).

10 Downing Street (Number Ten) the London residence of the Prime Minister (compare **11 Downing Street**).

1040 in US, the number (said 'ten-forty') of the personal income tax form (said jokingly to derive from the year 1040, when Lady Godiva rode naked through Coventry in protest against taxes, but in reality a simple serial number, introduced in 1913).

ten-four (10-4) in the **ten-code**, the signal for 'message received' or 'message correct'.

ten-gallon hat in US and Canada, a cowboy's broad-brimmed hat with a high crown (so called for its large size).

Ten Hours Act, the the act of 1847 forbidding children from working for more than ten hours a day.

ten-inch record a gramophone record (disc) 10 in in diameter, especially a **78** or LP (long-player) (compare **twelve-inch record**).

Ten Lost Tribes of Israel ten of the original **Twelve Tribes of Israel** which, led by Joseph, took possession of the Promised Land (Canaan) after the death of Moses: Asher, Dan, Ephraim, Gad, Issachar, Manasseh, Naphtali, Reuben, Simeon, Zebulun, all sons or grandsons of Jacob.

tenner in UK, a £10 note; in US, a $10 bill.

ten out of ten top marks, full score (sometimes used in its short form, 'ten', to express approval or admiration).

ten per center in US, a theatre agent (who takes 10 per cent commission).

tenpin bowling (tenpins) bowling with ten pins, the most common kind (with heavy bowls rolled down a long lane to knock down the pins at the end).

Ten Plagues of Egypt the plagues called down on Egypt by Jehovah through Moses to permit the Jews to escape from bondage: (1) water turned into blood, (2) plague of frogs, (3) of lice, (4) of flies, (5) of murrain, (6) of boils, (7) of hail, (8) of locusts, (9) of darkness, (10) death to the firstborn Egyptians (*Exodus* 7–12).

Ten Provinces the Dominion of Canada, as a federation of ten provinces and two territories: Alberta, British Columbia, Manitoba, New Brunswick, Newfoundland, Nova Scotia, Ontario, Prince Edward Island, Quebec, Saskatchewan, Northwest Territories, Yukon Territory.

'Ten Sixty-Six and All That' 'before my time', an expression denoting something that is antiquated or outdated (from title of humorous history of Britain by W C Sellar and R J Yeatman, 1930; 1066 was the year of the Norman invasion of Britain, a date that supposedly 'every schoolboy knows').

ten-strike in US, a success (from **tenpin bowling**, when such a strike knocks down all ten pins with the first bowl).

ten to one with chances strongly in favour.

ten-to-two with the hands or feet at the position of the hands of a clock at 1.50, for example when holding the steering wheel of a car when driving (in picture advertisements for clocks and watches, the time conventionally displayed is this one, supposedly because the clock face is then 'smiling').

ten-yard line in rugby, the line 10 yards (10m) from the halfway line, behind which the opposing players must stay when the ball is kicked off from the centre.

Ten Years War the unsuccessful revolution fought by Cuban guerrillas against the Spanish colonial regime from 1868 to 1878.

Top 10 the first ten best-selling pop records in the 'charts' (introduced in 1952 in UK from US, where first Top 10 was organized in 1942).

Upper Ten in US, the aristocracy or cream of society (short for 'the upper ten thousand', said to be the number of fashionable people resident in New York at the time when this phrase was coined in the 19th century by NPWillis, a US journalist).

Wilmington Ten the nine black men and one white woman imprisoned in 1972 after racial disturbances in Wilmington, North Carolina: Benjamin Clavis, Reginald Epps, Jerry Jacobs, James McKoy, Wayne Moor, Marvin Patrick, Connie Tindall, Anne Shephard Turner, Willi Earl Vereen, Joe Wright.

BASED ON TENTH

Tenth Avenue the New York Street associated with crime and violence, especially in the section between 40th and 55th Streets, called 'Hell's Kitchen'.

tenth muse a name given to the Greek poetess Sappho, in addition to the **Nine** (Muses), and later to various literary women, such as Mlle de Scudéry (in 17th century), Queen Christina of Sweden (18th century) and Hannah More (18th–19th centuries).

tenth wave the wave that is traditionally the largest in a running swell at sea (although Tennyson writes of the ninth wave thus in 'The Coming of Arthur' in *The Idylls of the King*, 1833).

QUOTATIONS AND SAYINGS

A dillar, a dollar,
A ten o'clock scholar,
What makes you come so soon?
You used to come at ten o'clock
And now you come at noon.
(nursery rhyme)

All lovers swear more performance than they are able, and yet reserve an ability that they never perform; vowing more than the perfection of ten, and discharging less than the tenth part of one. (Shakespeare, *Troilus and Cressida*, 1602)

My strength is as the strength of ten,
Because my heart is pure.
(Tennyson, 'Sir Galahad', 1842)

O, for ten years, that I may overwhelm
Myself in poesy.
(John Keats, 'Sleep and Poetry', 1817)

One man that has a mind and knows it can always beat ten men who haven't and don't. (Bernard Shaw, *The Apple Cart*, 1929)

Robin and Richard were two pretty men,
They lay in bed till the clock struck ten.
(nursery rhyme)

Roll up that map; it will not be wanted these ten years. (William Pitt, on a map of Europe, after hearing news of the Battle of Austerlitz, 1805)

Ten acres and a mule (what slaves sought in America from 1862, in their belief that their masters' plantations would be divided up to their benefit after the Civil War; compare 'Forty acres and a mule' under 40, and 'Three acres and a cow' under 3).

The good Lord has only ten. (Georges Clemenceau, referring to US President Woodrow Wilson's **Fourteen Points**; the 'ten' are the **Ten Commandments**)

There's a breathless hush in the Close to-night–
Ten to make and the match to win.
(Sir Henry Newbolt, *Vitaï Lampada*; the game is cricket)

The ten-year-old gives a fair indication of the man or woman he or she is to be. (Arnold Gesell, *The Child from Five to Ten*, 1946)

They are only ten. (Lord Northcliffe, newspaper proprietor; said to have been written up in his offices to remind the staff of the mental age of their readers)

ELEVEN

EXPRESSIONS

BASED ON ELEVEN

Chapter 11 in US, the provision of the Federal Bankruptcy Act for the relief of insolvent debtors and their creditors, by transferring the ownership of assets to a new entity owned by both creditors and debtors (compare **Chapter 7**, **Chapter 13**).

Eleven, the (1) a term sometimes used in the Bible for the 12 disciples (see the **Twelve**) without Judas Iscariot, for example in *Luke* 24:33: 'And they rose up the same hour . . . and found the eleven gathered together'; (2) in ancient Greece, the administrative body of 11 persons (ten elected by lot and their secretary) who had the duty of supervising prisons and prisoners and carrying out sentences (Greek name, *oi hendeka*, 'the eleven').

eleven days the 11 days dropped from the English calendar in 1752 when the change was made from the Julian calendar to the Gregorian, and 2 September was followed by 14.

11 Downing Street the London residence of the Chancellor of the Exchequer (compare **10 Downing Street**).

eleven-plus (11-plus) the former examination taken by children aged 11 or 12 in England and Wales to be selected for a grammar school; the system (and the name) still exists informally in some schools.

elevenses a mid-morning drink of tea or coffee, often with a snack, taken at about 11.00 a.m.

Eleven Years Tyranny the period 1629–40, when Charles I governed without summoning a parliament.

Number 11 see **11 Downing Street**.

BASED ON ELEVENTH

eleventh commandment 'Thou shalt not be found out' or 'Thou shalt not get caught' (a cynical or witty addition to the **Ten Commandments**).

eleventh hour the latest possible time, the final chance (from the biblical parable told in *Matthew* 20, where the vineyard labourers hired 'about the eleventh hour' were paid as much as those who had been working from early morning; the 'eleventh hour' would have been 5 p.m.).

QUOTATIONS AND SAYINGS

Give us back our eleven days! (popular slogan at the time of the **eleven days**, when many people believed they had been 'robbed' of calendar time)

I'm a second eleven sort of chap. (J M Barrie, *The Admirable Crichton*; the reference is to an inferior player, one not in a 'first eleven', or top cricket team)

TWELVE

EXPRESSIONS

BASED ON TWELVE

12 a film category (introduced 1989) denoting that the film has been passed for viewing only by persons aged 12 or over (compare **15**, **18**).

Twelve, the (1) the 12 disciples (apostles) of Jesus: Peter, Andrew, James and John (sons of Zebedee), Philip, Bartholomew, Thomas, Matthew, James (son of Alphaeus), Judas (Jude), Simon, Judas Iscariot (compare the **Eleven** (1)); (2) the 12 countries as members of the European Economic Community (Common Market) since 1986: Belgium, France, Italy, Luxembourg, Netherlands, West Germany, Denmark, Ireland, UK, Greece, Spain, Portugal (compare the **Six**, the **Seven** (2), the **Ten** (1)); (3) the Twelve Apostles of the Mormon Church, who as a Council are one of its 'General Authorities' (compare the **Seventy**).

twelve-bore a sporting gun with a diameter of spherical bullets at 12 to the pound.

Twelve Days of Christmas the 12 days from 26 December to 6 January (Epiphany).

twelve good men and true a term for an elected jury of 12 people (in England, but not in Scotland where there are 15) (they are 'true' as they bring a verdict, which word literally means 'true speech').

twelve-inch record a gramophone record with a diameter of 12 in, either a **78** or an LP (see **33**).

twelve-mile limit an offshore boundary 12 miles from the coast, claimed by some countries as the limit of their territorial jurisdiction.

twelvemonth a year (in poetic or archaic use).

Twelve Peers of France the principal warrior companions of Charlemagne, named in the *Chanson de Roland* as: Roland, Olivier, Gérin, Gérier, Bérengier, Otton, Samson, Engelier, Ivon, Ivoire, Anséis, Girard (French, *Les Douze Pairs*, or *Les Douze Paladins*).

twelve-pitch another name for élite, the typewriter typeface with 12 characters to the inch.

Twelvers, the a major Shi'ite sect of Islam,

believing in the succession of 12 imams, beginning with Ali ibn Abi Talib, the fourth caliph and son-in-law of Muhammad, and his 11 sons; the 12 imams are regarded as preservers of the faith, and pilgrimage to their tombs counts as a legitimate substitute for a pilgrimage to Mecca (see **Five Pillars of Islam**).

Twelve Tables the earliest code of Roman civil, criminal and religious law, promulgated in the 5th century BC; they were originally ten in number, with two added later, and were engraved on 12 tablets of bronze (or wood) (Latin, *Duodecim Tabulae*).

twelve-tone music the modern music invented and developed by Arnold Schoenberg from about 1915, and based on the 12 notes of the chromatic (twelve-note or dodecaphonic) scale.

Twelve Tribes of Israel the ancient name of the Hebrews, known as the Israelites because the tribes were named after the sons (or grandsons) of Jacob, whose name was changed by God to Israel: Reuben, Simeon, Judah, Issachar, Zebulun (sons of Jacob and his first wife Leah), Gad, Asher (sons of Jacob and Leah's maidservant, Zilpah), Dan, Naphtali (sons of Jacob and Bilhah, the maidservant of Rachel, Jacob's second wife), Manasseh, Ephraim (sons of Joseph, Jacob and Rachel's son, whose wife was Asenath), Benjamin (youngest son of Jacob and Rachel); all of these except Judah and Benjamin formed the **Ten Lost Tribes of Israel**, who settled in northern Palestine, while Judah and Benjamin were the progenitors of southern tribes.

Twelve Years Truce the truce of 1609 to 1621 in the **Eighty Years War** between Spain and the Netherlands.

BASED ON TWELFTH

Glorious Twelfth see (1) the **Twelfth**; (2) the **Twelfth of July**.

Twelfth (Glorious Twelfth) 12 August, when grouse shooting legally begins (see also **Twelfth of July**).

Twelfth Avenue a New York street formerly known as 'Death Avenue', for the youngsters killed by the trains on the tracks of the New York Central Railroad, which ran along it until 1938 (when the West Side Highway was constructed over 72nd Street).

Twelfth Day 6 January, the feast of the Epiphany.

twelfth man a reserve player for a cricket team (of 11).

Twelfth Night 5 January, the eve of the Epiphany; Shakespeare's play of the name was probably written to be performed on this day (perhaps with its first night in 1601), as a 'finale' to the **Twelve Days of Christmas**.

Twelfth of July the 'Glorious Twelfth', the anniversary of the Battle of the Boyne, actually 1 July (in the Julian calendar) 1690, when James II's Catholic army of French and Irish fled before William III's Protestant army of European mercenaries in Ireland.

'Twelfth Street Rag' the popular dance tune originally composed in 1909 by Euday Bowman, but popularized in the 1954 recording of it by Pee Wee Hunt.

QUOTATIONS AND SAYINGS

Nothing that happens after we are 12 matters very much. (J M Barrie, quoted by Andrew Birkin, *Sunday Times Magazine*, 8 October 1978).

The iron tongue of midnight hath told twelve;
Lovers, to bed; 'tis almost fairy time.
(Shakespeare, *A Midsummer Night's Dream*, 1595)

The jury, passing on the prisoner's life,
May in the sworn twelve have a thief or two.
(Shakespeare, *Measure for Measure*, 1604)

The twelve good rules, the royal game of goose. (Oliver Goldsmith, *The Deserted Village*, 1770; the 12 rules are those ascribed to Charles I: (1) urge no healths, (2) profane no divine ordinances, (3) touch no state matters, (4) reveal no secrets, (5) pick no quarrels, (6) make no comparisons, (7) maintain no ill opinions, (8) keep no bad company, (9) encourage no vice, (10) make no long meals, (11) repeat no grievances, (12) lay no wagers; goose was a game played with counters on a checkered board)

They took up of the fragments that remained twelve baskets full. (Feeding of the Five Thousand, Bible: *Matthew* 14:20)

'Why only twelve?' 'That's the original number.' 'Well, go out and get thousands.' (Samuel Goldwyn, attributed, during the filming of *The Last Supper*)

THIRTEEN

EXPRESSIONS

BASED ON THIRTEEN

Chapter 13 in US, the provision of the Federal Bankruptcy Act for the relief of insolvent debtors and their creditors, by relieving the debtor of the obligation to pay his creditors in full, or by giving an extension of time in which to pay, or both (compare **Chapter 7**, **Chapter 11**).

Old Thirteen the original 13 American colonies established by the British government in 1775: Georgia, Massachusetts, New Hampshire, New Jersey, New York, North Carolina, South Carolina, Virginia (the royal colonies), Delaware, Maryland, Pennsylvania (the proprietary colonies), Connecticut, Rhode Island (the corporate colonies) (compare the **Twenty-Four Proprietors**).

Thirteen Club a US association formed in the late 19th century to counter the popular association of 13 with ill luck, especially by arranging dinners with 13 present etc.

thirteen to the dozen very fast (a less common variant of **nineteen to the dozen**).

BASED ON THIRTEENTH

13th Amendment the amendment of the US Constitution that abolished slavery throughout the union.

Thirteenth Apostle St John Chrysostom, patriarch of Constantinople in the 4th and 5th centuries (compare the **Twelve**).

WRITER AS NAME AND NUMBER

The British soldier and writer T E Lawrence became famous as 'Lawrence of Arabia', a nickname given him by the American journalist Lowell Thomas. In order to escape the unwelcome attention that this sobriquet attracted, Lawrence entered the Royal Air Force as 352087 A/C John Hume Ross. The *Daily Express* rumbled his identity, however, so he was obliged to leave the RAF and enter the Tank Corps as 7875698 Pte T E Shaw. Eventually, he returned as an aircraftman to the RAF, where this time he became 338171 A/C Shaw. Noël Coward wrote to him while he was still an airman, and began his letter: 'Dear 338171 (May I call you 338?)' When Lawrence's account of life in the RAF was published, posthumously, under the title of *The Mint*, the name of its author was given on the cover as 'By 352087 A/C Ross'.

QUOTATIONS AND SAYINGS

If thirteen sit down to sup
And thou first have risen up,
Goodman, turn thy money!

(Sebastian Evans, *Brother Fabian's Manuscripts; and Other Poems*, 1865)

Years he number'd scarce thirteen
When fates turned cruel,
Yet three fill'd Zodiacs had he been
The stage's jewel.

(Ben Jonson, 'An Epitaph on Salomon Pavy, a Child of Queen Elizabeth's Chapel', *Epigrams*, 1672)

FOURTEEN

EXPRESSIONS

BASED ON FOURTEEN

fourteen-day rule an unofficial ruling in the Second World War (but later official from 1955 to 1957, when abolished) banning the discussion on radio or television of any issue due to come up in Parliament within the next fortnight.

Fourteen Holy Helpers the 14 saints popular in the Rhineland from the 14th century for their alleged power to cure diseases; the list of names varies, but traditionally is: Acacius, Barbara, Blaise, Catharine of Alexandria, Christopher, Cyriacus, Denys, Erasmus, Eustace, George, Giles, Margaret of Antioch, Pantaleon, Vitus (a painting of them by Cranach is at Hampton Court).

Fourteen Points US President Woodrow Wilson's pronouncement of 1918 of the war aims of the Allies in the First World War: (1) open diplomacy, (2) freedom of the seas, (3) freedom of trade, (4) general disarmament, (5) the impartial settlement of colonial claims, (6) a sympathetic treatment of the Russians, (7) the restoration of Belgium, (8) the return of Alsace-Lorraine to France, (9) a readjustment of the Italian frontiers, (10) autonomy for Austria-Hungary, (11) the restoration of Romania, Serbia and Montenegro, (12) autonomy for the Ottoman Empire, (13) an independent Poland, (14) the establishment of a general league of nations to promote peace and to guarantee every state's territorial integrity.

BASED ON FOURTEENTH

14th Amendment in US, the amendment to the

US Constitution that forbade any state from denying or abridging the federal rights of the citizens or their right to equal protection under the law.

Fourteenth of July in France, Bastille Day, a public holiday marking the Storming of the Bastille on 14 July 1789, the start of the French Revolution (French, *le Quatorze Juillet*).

Fourteenth Street in New York, the street regarded as a barrier between Downtown New York City and Midtown Manhattan, and a workers' district, where the Labor Forum is located as a meeting place for workers and trade unionists.

QUOTATIONS AND SAYINGS

All the generations from Abraham to David are fourteen generations; and from David until the carrying away into Babylon are fourteen generations; and from the carrying away into Babylon unto Christ are fourteen generations. (Bible: *Matthew* 1:17)

And they kept the passover on the fourteenth day of the first month at even in the wilderness of Sinai: according to all that the Lord commanded Moses, so did the children of Israel. (Bible: *Numbers* 9:5).

FIFTEEN

EXPRESSIONS

BASED ON **FIFTEEN**

Commission of Fifteen in US, the group appointed in the early 20th century to work out a compromise between the conservative and liberal elements in the Presbyterian church.

15 a film category, denoting that the film has been passed for viewing only by persons aged 15 or over (compare **12**, **18**).

Fifteen, the (1) the Jacobite rising of 1715, when James Edward Stuart, the Old Pretender, made an unsuccessful attempt to gain the throne (compare the **Forty-Five**); (2) the former name for the Court of Session, the supreme civil tribunal of Scotland, which was established in 1532 and originally had 15 judges (now 19).

fifteen ball a game of pocket billiards similar to rotation, using 15 consecutively numbered object balls and a white cue ball.

Fifteen Years War the war of 1591 to 1606 between Austria and the Ottoman Turks; it blocked Turkish expansion into Europe and contributed to the development of Transylvania as a significant power in eastern Europe.

BASED ON **FIFTEENTH**

15th Amendment in US, the amendment to the US Constitution declaring that the right to vote would not be denied 'on account of race, color, or previous condition of servitude'.

QUOTATIONS AND SAYINGS

Fifteen Men on the Dead Man's Chest. (song sung by the pirates in RLStevenson's *Treasure Island*, 1881; Stevenson got 'The Dead Man's Chest' from Charles Kingsley's *At Last*, 1871, where it is given as the name of one of the Virgin Islands)

Here's to the maiden of bashful fifteen. (Sheridan, *The Rivals*, 1775)

In the future everyone will be famous for fifteen minutes. (Andy Warhol, *Andy Warhol's Exposures*)

'To appreciate heaven well,
'Tis good for a man to have some fifteen minutes of hell.
(Will Carleton, *Gone With a Handsomer Man*, 19th century)

SIXTEEN

EXPRESSIONS

BASED ON **SIXTEEN**

16mm (16 mill, 16 millimetre) the film gauge to which feature films are reduced (from **35mm**) for private hire and, in many countries, for television.

Sixteen-String Jack a nickname of John Rann, a highwayman hanged in 1774, noted for his foppery (he wore 16 tags, eight at each knee)

sweet sixteen the conventional age of the 'bloom of youth' of a girl.

sweet sixteen party in US, a party given by the friends of a girl on her 16th birthday.

BASED ON **SIXTEENTH**

16th Amendment in US, the amendment to the US Constitution which (in 1913) introduced Federal Income Tax.

sixteenth note in US and Canada, the musical equivalent of a semiquaver (one-sixteenth of a semibreve).

QUOTATIONS AND SAYINGS

Age? Sixteen. The very flower of youth. (Latin, *Anni? sedecim. flos ipsus.*) (Terence, *Eunuchus I*, 161 BC)

I am sixteen, turning seventeen, innocent as the day. (song in film *The Sound of Music*, 1965)

Now, by the bless'd Paphian queen,
Who heaves the breast of sweet sixteen.
(Oliver Wendell Holmes, *The Dilemma*, 1840)

SEVENTEEN

EXPRESSIONS

BASED ON **SEVENTEEN**

sweet seventeen 'used typically for the most attractive period of a girl's life'. (*Oxford English Dictionary*)

BASED ON **SEVENTEENTH**

17th Amendment in US, the amendment to the US Constitution (in 1913) that Senators should be elected by popular vote.

17th Congress in USSR, the Communist Party Congress of 1934, in which Stalin admitted that the peasantry had slaughtered half the country's livestock as a protest against collectivization.

17th Parallel the parallel or boundary (17°N) fixed in 1954 between North and South Vietnam.

QUOTATIONS AND SAYINGS

Claspt hands and that petitionary grace
Of sweet seventeen subdued me ere she spoke.
(Tennyson, *The Brook*, 1855)

I'm seventeen come Sunday. (old popular song)

Sweet seventeen – and never been kissed. (popular saying)

EIGHTEEN

EXPRESSIONS

BASED ON **EIGHTEEN**

18 a film category, denoting that a film has been passed for viewing only by persons aged 18 or over (compare **12**, **15**).

18-8 stainless steel steel that contains about 18 per cent chromium and 8 per cent nickel, developed in Germany.

BASED ON **EIGHTEENTH**

18th Amendment in US, the amendment to the US Constitution (of 1919) introducing Prohibition, which came into force in 1920.

QUOTATIONS AND SAYINGS

18 Adulthood begins. You can: vote, be on a jury in England and Wales, buy a house, see an X film, have a cheque or credit card, bet, drink alcohol in a pub, or buy it from an off-licence, be liable for overdrafts now incurred, make a will, sign documents (like insurance policies) for yourself, get married without parents' consent in England and Wales. (*Which?* January 1974; the X film certificate has now been superseded by the **18** category)

From birth to age eighteen, a girl needs good parents. From eighteen to thirty-five, she needs good looks. (Sophie Tucker, said when she was 69)

NINETEEN

EXPRESSIONS

BASED ON **NINETEEN**

Nineteen Propositions the 19 points in the Long Parliament's demands of Charles I on 1 June 1642; they were rejected by him, and were the immediate cause of the English Civil War; they included demands that parliamentary approval should be required for the appointment and dismissal of all officers of state, that the church should be reformed on lines directed by parliament, and that parliaments should control the militia.

nineteen to the dozen very fast, especially of speech (perhaps 'nineteen' as the highest of the teens).

BASED ON **NINETEENTH**

19th Amendment in US, the amendment to the US Constitution (of 1920) which introduced women's suffrage.

nineteenth hole the bar of a golf club (as a favourite resort after completing the 18 holes of the course; players 'sink' their drinks as they 'sink' their putts).

QUOTATIONS AND SAYINGS

'How old are you, Sally?' 'Nineteen.' 'Good God! And I thought you were about twenty-five!' 'I know. Everybody does.' (Sally Bowles in Christopher Isherwood's *Goodbye to Berlin*, 1939)

Would any but these boiled brains of nineteen and two-and-twenty hunt this weather? (Shakespeare, *Measure for Measure*, 1604)

TWENTY

EXPRESSIONS

BASED ON **TWENTY**

Roaring Twenties the 1920s, a decade of youthful uninhibitedness, 'flappers' and 'living it up', mainly as a reaction to the austerities of the First World War (probably an expression based on **Roaring Forties**; compare **Swinging Sixties**).

Society of the Twenty the official title of a group of 20 artists who exhibited together in Belgium in 1891–3, with a common interest in Symbolism (French, *Les Vingt*).

twenty questions a formerly popular parlour game, taken up on radio, in which a person has to guess the identity of someone or something by asking up to 20 questions, which can be answered by only 'yes' or 'no'.

twenty-to-one (20-1) in horseracing, tipped to win.

twenty-twenty (20/20) vision perfect eyesight (officially the Snellen fraction to express characters read perfectly clearly at a distance of 20ft; the repeated '20' refers to the two eyes).

BASED ON **TWENTIETH**

twentieth-century cut in diamond cutting, having 80 or 88 facets, a cut introduced at the turn of the 20th century.

Twentieth Century Limited the same of an express train that ran between Chicago and New York from 1902 to 1967.

20th Congress in USSR, the 20th Communist Party Congress of 1956, when Khrushchev denounced Stalin in his 'secret speech' and announced a reversal of policy.

QUOTATIONS AND SAYINGS

A club is a place where twenty men pay for the pleasure of one. (anonymous saying)

Annual income twenty pounds, annual expenditure nineteen nineteen six, result happiness. Annual income twenty pounds, annual expenditure twenty pounds ought and six, result misery. (Mr Micawber, in Charles Dickens', *David Copperfield*, 1849)

How old is twenty? Ask Ten: ask Thirty: ask Einstein. (Ralph Hodgson, *Flying Scrolls*, 1961)

Live as long as you may, the first twenty years are the longest half of your life. (Robert Southey, *The Doctor*, 1934–47)

Now of my threescore years and ten,
Twenty will not come again,
And take from seventy springs a score,
It only leaves me fifty more.
(A E Housman, *A Shropshire Lad*, 1896)

Then come kiss me, sweet and twenty,
Youth's a stuff will not endure.
(Shakespeare, *Twelfth Night*, 1601: popularly taken to refer to a 'sweet girl of 20', but 'and twenty' is merely an intensive, meaning 'very', as if 'twenty times sweet')

Twenty love-sick maidens we,
Love-sick all against our will.
(Gilbert and Sullivan, *Patience*, 1881)

TWENTY-ONE

EXPRESSIONS

BASED ON **TWENTY-ONE**

twenty-one pontoon, *vingt-et-un*, a gambling game with cards in which the players try to beat the banker but not score a card combination over 21 points.

Twenty-One Demands the 21 demands made in 1915 by Japan on China with the aim of increasing their rights in Shantung, Mongolia and Manchuria.

twenty-one gun salute in UK, a royal salute of 21 guns fired ceremoniously; in US, a similar presidential salute (21 guns was the largest number formerly found along one side of a 'ship of the line'; the firing of all these was thus the highest mark of respect; the US ceremony was adopted from the British navy in 1875; the US President is accorded the salute as Commander-in-Chief).

Twenty-One Years War the war of 1701 to 1721 between Russia and Sweden, in which Russia gained access to the Baltic.

BASED ON **TWENTY-FIRST**

twenty-first(er) before 1970, a birthday party

given to celebrate a person's coming of age at 21, and still held even now.

21st Amendment in US, the amendment to the US Constitution (of 1933) that repealed Prohibition.

twenty-first (21st) century the future, and increasingly the *near* future.

QUOTATIONS AND SAYINGS

But I was one-and-twenty,
No use to talk to me.
(A E Housman, *A Shropshire Lad*, 1896)

I'm twenty-one today,
Twenty-one today!
I've got the key of the door,
Never been twenty-one before!
(old popular song, celebrating a coming of age)

TWENTY-TWO

EXPRESSIONS

BASED ON TWENTY-TWO

catch 22 a dilemma from which it is impossible to escape, as both alternatives are mutually exclusive (from title of novel by Joseph Heller, 1961; in this, the reference is to a fictional rule of the US Air Force in the Second World War: a pilot must continue flying combat missions until he is judged insane or otherwise unfit; if he continues flying into anti-aircraft fire, he is probably insane; but if he asks to be relieved, he is probably sane, and may not be relieved; either way, he keeps on flying until he meets his inevitable fate; Heller originally called his book *Catch 18*, but was obliged to change it when Leon Uris's *Mila-18* was published shortly before his own book was due out).

twenty-two (1) in US, a **two-two** rifle; (2) in rugby, a line across the pitch 22 m from the goal (compare **twenty-five**).

twenty-two carat the same as **twenty-four carat**.

BASED ON TWENTY-SECOND

22nd Amendment in US, the amendment to the US Constitution (in 1951) which limited the tenure of the Presidency to two terms.

QUOTATIONS AND SAYINGS

The myrtle and ivy of sweet two-and-twenty
Are worth all your laurels, though ever so plenty.
(Byron, *Stanzas Written on the Road between Florence and Pisa*, 1821).

TWENTY-THREE

EXPRESSIONS

BASED ON TWENTY-THREE

twenty-three skidoo 'go away', 'scram' (a phrase popular in the 1920s and 1930s, and said to refer to 23rd Street, New York, from which many railroads left the city).

BASED ON TWENTY-THIRD

Twenty-Third (23rd) Psalm one of the most popular and most familiar of the Psalms, set to music in various forms, and beginning 'The Lord is my shepherd . . .'

QUOTATIONS AND SAYINGS

Lord, how ashamed I should be of not being married before three and twenty! (Lydia, in Jane Austen's *Pride and Prejudice*, 1813)

She
Was married, charming, chaste, and twenty-three.
(Byron, *Don Juan*, 1819–24)

TWENTY-FOUR

EXPRESSIONS

24 Hours (of Le Mans) the round-the-clock motor race at Le Mans, France (French; *Les Vingt-Quatre Heures du Mans*).

twenty-four (24) carat (1) entirely, thoroughly; (2) genuine, trustworthy (both senses from the proportion of gold in an alloy, with 24 being the equivalent of 100 per cent).

twenty-four (24)-hour lasting day and night.

twenty-four (24)-hour clock a clock, or statement of the time, which reckons to 24, not 12, so that 1 a.m. is 0100 and 1 p.m. is 1300.

Twenty-Four Proprietors, the the 24 Quakers (William Penn and his associates) who bought East Jersey (present state of New Jersey) in 1682.

QUOTATIONS AND SAYINGS

Four-and-twenty blackbirds,
Baked in a pie
(from nursery rhyme 'Sing a song of sixpence'; the 24 birds have been explained variously, as the hours of the day, the choirs of monasteries about to be dissolved, etc.)

Four and twenty tailors went to kill a snail. (nursery rhyme, with various explanations for the figure)

Pitt was a Prime Minister at four-and-twenty, and that precedent has ruined half our politicians. (Anthony Trollope, *Phineas Finn*, 1869)

There are only 24 hours in a day. (saying)

TWENTY-FIVE

EXPRESSIONS

twenty-five, (25)-yard line the line on a hockey pitch (and formerly on a rugby pitch, see **twenty-two**) 25 yards from the goal line.

Twenty-Five Articles of Religion the articles prepared by John Wesley for the Methodist Church in the US in 1784, and basically an abridgement of the **Thirty-Nine Articles** of the Church of England.

QUOTATIONS AND SAYINGS

The first twenty-five years of one's life are worth all the rest of the longest life of man, even though those five-and-twenty be spent in penury and contempt, and the rest in the possession of wealth, honours, respectability. (George Borrow, *The Romany Rye*, 1857)

Twenty-five seems to me to be the ideal age at which to write one's autobiography. (Beverley Nichols, *Twenty-Five*, 1926)

TWENTY-SIX

EXPRESSIONS

twenty-six a dice game in which a player selects a number from 1 to 6 and then throws 10 dice 13 times, aiming to throw that number 26 times or more.

Twenty-Six Counties the counties of the Irish Free State, formed in 1921 and now the Republic of Ireland: Galway, Leitrim, Mayo, Roscommon, Sligo (in Connacht), Carlow, Dublin, Kildare, Kilkenny, Laoighis, Longford, Louth, Meath, Offaly, Westmeath, Wexford, Wicklow (in Leinster), Clare, Cork, Kerry, Limerick, Tipperary (in Munster), Cavan, Donegal, Monaghan (in Ulster) (compare the **Six Counties**, and see also the **Five Fifths of Ireland**).

QUOTATIONS AND SAYINGS

It was not long after that that everybody was twenty-six. During the next two or three years all the young men were twenty-six years old. It was the right age apparently for that time and place. (Gertrude Stein, *The Autobiography of Alice B. Toklas*, 1933)

Towards the age of twenty-six
They shoved him into politics.
(Hilaire Belloc, *Cautionary Tales*, 1907)

TWENTY-SEVEN

EXPRESSIONS

27 Club at Harrow School, a club for political discussion (founded in 1927).

QUOTATIONS AND SAYINGS

It was about then [1920] that I wrote a line which certain people will not let me forget: 'She was a faded but still lovely woman of twenty-seven'. (Scott Fitzgerald, *Early Success*, 1937)

TWENTY-EIGHT

EXPRESSIONS

Clause 28 in the Local Government Act of 1988, the clause that prohibited the promotion of homosexuality by local authorities, and that was therefore contentious.

QUOTATIONS AND SAYINGS

Girls who don't marry by twenty-eight usually stay at that age until they do. (anonymous quotation)

TWENTY-NINE

EXPRESSIONS

twenty-niner a person who was born on 29

February, so who can mark their birthday accurately only in a leap year.

QUOTATIONS AND SAYINGS

Nine-and-twenty knights of fame
Hung their shields in Branksome Hall;
Nine-and-twenty squires of name
Brought them their steeds to bower from stall;
Nine-and-twenty yeomen tall
Waited, duteous, on them all.
(Walter Scott, *The Lay of the Last Minstrel*, 1805)

The best ten years of a woman's life are between 29 and 30. (anonymous saying)

THIRTY

EXPRESSIONS

Hungry Thirties the 1930s, when there was much unemployment and related 'hunger marches', such as the famous one of 1936 from Jarrow to London (compare **Hungry Forties**).

like thirty cents in US, cheap, worthless.

thirty (30) in journalism, 'that's all', 'the end', on completion of a message or report (from the early telegraphic use of 'XXX' at the end of a story).

thirty minutes in US, a call to actors that the show begins in half an hour.

30 Rock in US, the abbreviated address of 30 Rockefeller Plaza, the New York headquarters of NBC-TV.

Thirty Tyrants (1) the 30 magistrates appointed by Sparta over Athens at the end of the Peloponnesian war (404 BC); (2) in the Roman Empire, the military usurpers who endeavoured in the 3rd century to make themselves independent princes (although only 19 are numbered, so '30' is inexact; they were named after their ancient Greek counterparts).

thirty-year rule the rule that public (especially government) records are open to inspection only 30 years after their compilation; this results in disclosures of previously confidential material being made every New Year.

Thirty Years War the major European war of 1618–48, beginning between the Catholics of South Germany and the Protestants of North Germany.

QUOTATIONS AND SAYINGS

For a young man, a woman of thirty has irresistible attractions. (Balzac, *The Woman of Thirty*, 1832)

They covenanted with him for thirty pieces of silver. (Bible: *Matthew* 26:15; the sum for which Judas Iscariot betrayed Jesus with a kiss; 30 pieces of silver [probably Greek tetradrachms] was the value of a slave if killed by a beast; in the Old Testament, Joseph was sold for 20 pieces of silver [*Genesis* 37:28], or two-thirds the price of an adult slave)

Thirty days hath November,
April, June, and September,
All the rest have thirty-one
Excepting February alone.
(traditional rhyme dating back to at least 16th century, and having several variations)

Thirty – the promise of a decade of loneliness, a thinning list of single men to know, a thinning briefcase of enthusiasm, thinning hair. (Scott Fitzgerald, *The Great Gatsby*, 1925)

31 TO 39

EXPRESSIONS

thirty-one *trente-et-un*, a game of cards in which the aim is to hold three cards of the same suit scoring 31 (the ace counting 11 and the court cards 10 each).

thirty-three (33) a long-playing record turning at 33⅓ revolutions a minute (compare **45**, **78**).

thirty-five millimetre (35 mill, 35 mm) the standard size of roll or movie film (compare **16 mm**).

38th Parallel the boundary between North and South Korea, established in 1945 (and running along the line of latitude 38°N).

Thirty-Nine Articles in the Christian religion, the doctrinal statement drawn up by the Church of England in the second half of the 16th century as a restatement of Archbishop Cranmer's **Forty-Two Articles**, to which all Anglican clergy must assent on their ordination (the number has been jokingly referred to by some theological students as 'Forty stripes save one').

QUOTATIONS AND SAYINGS

I shall be thirty-one . . . My youth is like a dream; and very little use have I ever made of it. What have I done these last thirty years? Precious little. (Charlotte Brontë, letter to Ellen Nussey, 1847)

I am thirty-three – the age of the good *sans-culotte* Jesus; an age fatal to revolutionists. (Camille Desmoulins, when asked his age by the French Revolutionary Tribunal, 1794; his real age was 34; he was guillotined two days later)

Thirty-five is a very attractive age. London society is full of women of the very highest birth who have, of their own free choice, remained thirty-five for years. (Lady Bracknell in Oscar Wilde's play, *The Importance of Being Earnest*, 1895)

FORTY

EXPRESSIONS

Ali Baba and the Forty Thieves a famous story in the **Thousand and One Nights** (the number is arbitrary).

Field of the Forty Footsteps the land behind the British Museum, London, formerly called Southampton Fields (said to have been named from the footprints left by two brothers, when they duelled to the death in 1685).

Forty, the the 40 members of the French Academy (French *Les Quarante*).

Forty Hours in the Roman Catholic church, the devotion in which the Blessed Sacrament is exposed for about 40 hours, with the faithful praying before it.

forty-legs a dialect word for a centipede ('forty' meaning simply 'many').

Forty Martyrs (of England and Wales) the 40 Roman Catholics put to death by the state between 1535 and 1680, selected as martyrs from the 200 already beatified by earlier popes; they were canonized in 1970 and include the **Three Jesuit Martyrs** – Edmund Campion, Alexander Briant and Ralph Sherwin; the other 37 were: John Almond, Edmund Arrowsmith, Ambrose Barlow, John Boste, Margaret Clitherow, Philip Evans, Thomas Garnet, Edmund Gennings, Richard Gwyn, John Houghton, Philip Howard, John Jones, John Kemble, Luke Kirby, Robert Lawrence, David Lewis, Anne Line, John Lloyd, Cuthbert Mayne, Henry Morse, Nicholas Owen, John Paine, Polydore Plasden, John Plessington, Richard Reynolds, John Rigby, John Roberts, Alban Roe, Robert Southwell, John Southworth, John Stone, John Wall, Henry Walpole, Margaret Ward, Augustine Webster, Swithun Wells, Eustace White.

Forty Martyrs of Sebaste the 40 Christian soldiers of the Roman 'Thundering Legion' (*Lcgio XII Fulminata*) martyred at Sebaste (modern Sivas, Turkey) in about AD 320.

forty-rod lightning in US, whiskey (from its strength, which could 'kill at 40 yards' like a lightning strike).

forty winks a brief nap, especially one after dinner (simply an arbitrary number; nappers wink or blink more than sound sleepers).

Hungry Forties the late 1840s, with much poverty as a result of poor harvests, dear bread, and unemployment (compare **Hungry Thirties**).

Roaring Forties the stormy areas of ocean between 40° and 50° in the southern hemisphere.

Room 40 the British government's military decoding centre in the Second World War.

QUOTATIONS AND SAYINGS

fair, fat and forty (a description of a comely but buxom woman; perhaps adopted from John

O'Keeffe, *Irish Mimic*, 1795, where the original runs: 'Fat, fair and forty were all the toasts of the young men')

Forty acres and a mule. (an expansion of the slogan 'Ten acres and a mule', which see under 10; the expanded slogan arose in 1865, when General Sherman stated that 'Every family shall have a plot of not more than forty acres of tillable land'; compare **Three acres and a cow** under 3)

Forty days and forty nights. (the length of time that Jesus fasted in the wilderness, in *Matthew* 4:2: 'And when he had fasted forty days and forty nights, he was afterward an hungred'; the number is arbitrary, meaning 'a long time')

Forty stripes save one. (that is, 39, with the full quotation in *II Corinthians* 11:24, where St Paul writes: 'Of the Jews five times received I forty stripes save one'; the Jews were forbidden by Mosaic law to inflict more than 40 stripes on an offender, so stopped short of this number; if a scourge had three lashes, 13 strokes would be enough to give 'forty stripes save one')

Forty years old is not yet a woman;
Forty degrees below is not yet a frost.
(Siberian proverb)

Forty years on (the opening words, and title, of the Harrow School Song) (see box, p. 90)

I'll put a girdle round the earth
In forty minutes.
(Puck, in Shakespeare, *A Midsummer Night's Dream*, 1595)

Life begins at forty. (in other words: the best years are yet to come; the saying was popularized by a song of the name in the 1940s, written by Jack Yellen for Sophie Tucker)

Lizzie Borden took an ax
And gave her mother forty whacks;
When she saw what she had done
She gave her father forty-one!
(anonymous rhyme after the trial of Lizzie Borden at Fall River, Massachusetts, in 1893)

41 TO 49

EXPRESSIONS

Centre 42 Arnold Wesker's projected organization of artists and writers to provide a cultural entertainment for the working classes at the Round House Theatre, London, in 1960; the venture was not a success, and was disbanded in 1970 (the number refers to Resolution 42 of the Trades Union Congress in 1960, calling on the TUC to participate more fully in the arts).

42nd Street the New York street associated first with vaudeville theatres, then with honky-tonk, 'adult' movies and amusement arcades, with now a general air of 'sleaziness'.

Forty-Two Articles the doctrine of the Church of England devised by Archbishop Cranmer and issued in 1553; it was never enforced, as the Roman Catholic faith was restored under Mary I (compare **Thirty-Nine Articles**).

Rule 43 the prison rule allowing segregation of prisoners for their own protection.

forty-five (45) (1) an EP (extended play) 7-inch record, turning at 45 revolutions a minute (compare **33**, **78**); (2) in US, a pistol with a calibre of .45 inches.

Forty-Five, the the (2nd) Jacobite rebellion, of 1745, led by Charles Edward Stewart, the Young Pretender, but crushed at Culloden in 1746.

47th Street the New York street between **Fifth Avenue** and Sixth Avenue, as the location of the diamond market.

47 Workshop in US, the experimental dramatic class set up by GPBaker in 1905 at Harvard University, and running for 20 years (named after the course number).

48 in the services, a pass for a weekend leave (of 48 hours, Saturday and Sunday).

Forty-Eight, the the short title of Bach's collection of 48 preludes and fugues, which he called *The Well-Tempered Clavier*, 1722 and 1742; the work contains a prelude and a fugue in each of the 12 major and 12 minor keys.

Forty-Niner a participant in the Californian gold rush of 1849, after it had been discovered at Sutter's Mill the previous year.

49th Parallel the boundary between the USA and Canada, from the Pacific coast to a point south of Winnipeg (it runs along the line of latitude 49°N).

QUOTATIONS

Mr Salteena was an elderly man of 42 and was fond of asking peaple (*sic*) to stay with him. (opening words of Daisy Ashford, *The Young Visiters*, 1919)

She may very well pass for forty-three

In the dusk with a light behind her!
(Gilbert and Sullivan, *Trial by Jury*, 1875)

She's six-and-forty, and I wish nothing worse to happen to any woman. (Sir Arthur Wing Pinero, *The Second Mrs Tanqueray*, 1893)

When you are 48 years old you have done enough. (Ben Abruzzo, after being the first man, in 1978, to cross the Atlantic by balloon)

The body is at its best between the ages of thirty and thirty-five: the mind at its best about the age of forty-nine. (Aristotle, *Rhetoric*, 4th century BC)

50 TO 59

EXPRESSIONS

Feeble Fifty a nickname for the 50 Labour MPs in the 1980s (to 1988, when they became 49), as being ineffective in mitigating the advance of Thatcherism.

fifty-fifty (50-50) equal, half-and-half.

Committee of Fifty One in US, the grouping formed in New York in 1774, as favouring a general congress.

Fifty-First State in US, a nickname for Puerto Rico, as a commonwealth (from 1952) whose citizens have US citizenship ('51st' from 1959, when Hawaii became the 50th state).

Fifty-Two Per Cent the female population of the world in the 1980s.

56 in the services, a three-day leave period, especially for someone who has been on duty the previous weekend.

QUOTATIONS AND SAYINGS

Love is lame at fifty years. (Thomas Hardy, 'The Revisitation', *Time's Laughingstocks*, 1909)

Half-owre, half-owre to Aberdour,
'Tis fifty fathoms deep
(ballad, *Sir Patrick Spens*)

Dread fifty above more than fifty below. (Robert Frost, *Goodbye and Keep Cold*, 1923)

Fifty-four forty or fight. (the expansionist slogan of the Democrats in the US presidential election campaign of 1844, when war with Britain seemed imminent; the campaign was won by President Polk, who wanted to exclude Britain from the Oregon Territory, whose northern boundary, at 54° 40'N, was contiguous with Alaska; however, Polk and the Senate compromised, and agreed to the British proposal that the boundary between Canada and the US should be continued along the **49th Parallel**)

60 TO 69

EXPRESSIONS

like sixty in US, very fast.

LX Club the Cambridge University rugby club of the best 60 players in residence, elected for life.

Swinging Sixties the 1960s, by contrast with the austere years following the Second World War (compare **Roaring Twenties**).

sixty-four (64) dollar question the most difficult question (from a US radio quiz show of the 1940s, a 'double-or-quits' show, which had $64 as the highest prize) (compare **sixty-four thousand dollar question**).

Route 66 in US, the main road between Chicago and Los Angeles before the construction of the interstate highway system (made famous in the song 'Get Your Kicks On Route 66' by Bob Troup, 1946).

sixty-nine (more usually, **soixante-neuf**) as a sexual activity, mutual cunnilingus and fellatio (with the position of the bodies representing the figures of '69').

QUOTATIONS

If you can fill the unforgiving minute
With sixty seconds' worth of distance run,
Yours is the Earth and everything that's in it.
(Rudyard Kipling, *If–*, 1910)

I'm sixty-one today,
A year beyond the barrier,
And what was once a Magic Flute
Is now a Water Carrier.
(anonymous verse)

Will you still need me,
Will you still feed me,
When I'm sixty four?
(Lennon and McCartney, 1967)

I'm 65 and I guess that puts me in with the geriatrics, but if there were 15 months in every year, I'd only be 48. (James Thurber)

70 TO 79

EXPRESSIONS

Seventy, the (1) the body of disciples

appointed by Christ to preach the gospel and heal the sick (*Luke* 10:1); (2) the 'seventy interpreters' who are traditionally said to have translated the Septuagint, the Greek version of the Old Testament; (3) the First Council of Seventy, or 70 Mormons who rank in the hierarchy after the Twelve Apostles (see **Twelve** (3)), by whom they are directed, and who have the duty 'to travel into all the world and preach the gospel'.

seventy-three in US, 'best wishes', 'goodbye' (taken arbitrarily from Morse signals, where, for example, 4 means 'where' and 22 means 'kisses').

seventy-five a 75-mm gun, used by the US and France in the First World War.

Charter 77 a document signed by over 200 Czech citizens in 1977, protesting against their government's failure to observe international agreements on human rights.

seventy-eight (78) a gramophone record of pre-LP vintage, turning at 78 revolutions a minute (and abandoned by the end of the 1950s) (compare **33**, **45**).

QUOTATIONS AND SAYINGS

Seventy is the threshold of middle age. (Judge Melford Stevenson, who retired in 1979 at the age of 76)

The days of our age are threescore years and ten. (Book of Common Prayer)

My diseases are an asthma and a dropsy and what is less curable, seventy-five. (Samuel Johnson, letter to W G Hamilton, 1784)

At seventy-seven it is time to be in earnest. (Boswell's *Life of Johnson*, 1786)

80 TO 89

EXPRESSIONS

Eighty Years War the war of 1568 to 1648, when the Netherlands fought for independence from Spain (and won).

eighty-six in US, to refuse to serve a customer, to ban or bar or eject (probably as number rhymes with 'nix', otherwise nothing, referring to the inability of a customer to pay).

eighty-eight in US, a piano (from the 88 keys of a standard piano).

QUOTATIONS

At seventeen years many their fortunes seek;
But at fourscore it is too late a week.
(Shakespeare, *As You Like It*, 1599)

Toothless eighty. (Ebenezer Elliot, *Poems*, 1835)

You couldn't live eighty-two years in the world without being disillusioned. (Rebecca West, 1975)

Forty and forty-five are bad enough; fifty is simply hell to face; fifteen minutes after that you are sixty; and then in ten minutes more you are eighty-five. (Don Marquis)

90 TO 99

EXPRESSIONS

Naughty Nineties the 1890s, with their lax morality and light-hearted pleasures, expressed through the music-hall, as a reaction to earlier Victorian prudery and puritanism.

ninety-day wonder in US services, a graduate of a 90-day officers' training course in the Second World War, so one who was inexperienced or newly-commissioned (based on **nine days wonder**).

Ninety-Five Theses the propositions for a debate on indulgences, written in Latin and posted on the door of the Schlosskirche, Wittenberg (now East Germany) by Martin Luther in 1517; this was the birth of the Protestant Reformation.

ninety-nine times out of a hundred very frequently, mostly.

say 'ninety-nine' formerly, a doctor's instruction to a patient when sounding his chest or back for possible pneumonia or related disease (the words resonate when heard through a stethoscope, but may have been suggested by a body temperature of 99°F, one slightly above average). The French equivalent is (or was) *trente-trois* (33) which, because of its nasal sounds, similarly resonates.

QUOTATIONS AND SAYINGS

Primates: Man (neolithic): Homo sapiens: 90 ('Life Spans of Selected Mammals, Birds, Reptiles, and Invertebrates', *Encyclopaedia Britannica*)

The first 90 minutes are the most important. (traditional exhortation by team manager to football team before match; the game lasts 90 minutes)

Every morning, when you are 93, you wake up and say to yourself 'What, again?' . . . I'll tell you the really great thing about living to be 93: one does not have any rivals, because they're all dead. (Ben Travers, in conversation with Alan Ayckbourn, 1979)

Joy shall be in heaven over one sinner that repenteth, more than over ninety and nine just persons, which need no repentance. (Bible: *Luke* 15:7)

Ninety-nine clomp. ('centipede with a wooden leg': children's humorous definition)

HUNDRED

EXPRESSIONS

Committee of 100 the executive committee of the CND (Campaign for Nuclear Disarmament), led by Bertrand Russell and comprising prominent people from show business, the arts and literature; members were sent to prison for their activities in 1961, and the organization was disbanded in 1968.

hundred a former county division.

hundred and one many, countless (compare **thousand and one**); the expression implies variety, and is common in book titles such as 'A Hundred and One Basic Recipes'.

Hundred Associates in Canada, a popular name for the Company of New France, a joint-stock company chartered in 1627.

Hundred Days (1) the second rule of Napoleon, from when he entered Paris on 20 March 1815 to his abdication on 22 June that year (after the Battle of Waterloo); (2) in US, the period 9 March to 16 June 1933, when Congress, called into special session by President FD Roosevelt, passed most of the reform and recovery laws for the New Deal program that he presented to them to end the Depression; (3) the transitory socialist government established in Chile from June to December 1932.

Hundred Flowers in China, the period of approximately six weeks in the summer of 1957, when certain elements of the population were invited to criticize the current Communist political system under Chairman Mao.

hundred-legs a dialect word for a centipede.

hundreds and thousands small coloured particles of sugar used for cake decoration.

hundredweight a weight of 112lb (50.7kg) (originally 100lb (45.3kg)).

Hundred Years War the intermittent war between England and France that ran from 1337 to 1453, in which English kings claimed the French crown.

not a hundred miles from quite close to.

Old Hundred (Old Hundredth) the hymn tune, first recorded in the 16th century, that came to be set to Psalm 100 in the 'old' metrical version of the Geneva Psalter; it is Hymn 166 in *Hymns Ancient and Modern* and begins with the words 'All people that on earth do dwell'.

one hundred per cent entirely, fully, often with reference to a person's health, as 'not quite one hundred per cent'.

QUOTATIONS

All this will not be finished in the first one hundred days. (President JFKennedy, Inaugural Address, 1961)

A simple maiden in her flower
Is worth a hundred coats-of-arms.
(Tennyson, *Lady Clara Vere de Vere*, 1832)

Had I a hundred tongues, a hundred lips, a throat of iron and a chest of brass, I could not tell men's countless sufferings. (Virgil, *Aeneid VI*, 1st century BC)

I have defined the hundred per cent American as ninety-nine per cent an idiot. (Bernard Shaw, *Remarks on Sinclair Lewis receiving the Nobel Prize*, 1930)

One father is more than a hundred schoolmasters. (George Herbert, *Jacula Prudentum*, 1651)

The buyer needs a hundred eyes, the seller not one. (George Herbert, *Jacula Prudentum*, 1651)

Who wants to live to be a hundred? What's the point of it? (Henry Miller, 'On Turning Eighty', 1971)

200 TO 900

EXPRESSIONS

Four Hundred in US, the most prestigious or affluent social group or clique in a place (from the remark made by Ward McAllister in the *New York Herald Tribune*, 1888, that 'There are only about four hundred people in New York society'; a satirical counterblast to this was O Henry's volume of short stories, *The Four Million*, 1906, implying that the whole population of New York, then 4 million, were the people who mattered).

Indianapolis 500 (Indy 500) in US, the motor race held annually since 1911 over a 500-mile-plus course at the Speedway, Indianapolis.

QUOTATIONS

O what a world of vile ill-favour'd faults
Looks handsome in three hundred pounds a year!
(Shakespeare, *The Merry Wives of Windsor*, 1597)

Five hundred a year! I am sure I cannot imagine how they will spend half of it. (Mrs J Dashwood, in Jane Austen's, *Sense and Sensibility*, 1811)

I often wish'd that I had clear,
For life, six hundred pounds a year
(Jonathan Swift, *Imitations of Horace*, 1738)

Seven hundred pounds and possibilities is goot gifts. (Shakespeare, *The Merry Wives of Windsor*, 1597)

THOUSANDS

EXPRESSIONS

death of a thousand cuts a series of minor hurts or annoyances that eventually cause major damage or harm.

Expedition of a Thousand Garibaldi's campaign of 1860 in which he overthrew the Bourbon kingdom of the **Two Sicilies**, taking with him (at least) 1000 men.

Land of a Thousand Lakes a nickname of Finland (which has nearer 60,000).

Mille Miglia in Italy, the former motor race of about 1000 miles in length from Brescia to Cremona and back; it was first held in 1927 and abandoned, after many fatalities, in 1957.

One Thousand Guineas the annual horserace at Newmarket for fillies, first held in 1814 (the prize being 1000 guineas, or £1050) (compare **Two Thousand Guineas**).

Thousand Days (1) the period of 4 January 1961 to 22 November 1963, served by US President John F. Kennedy in office; (2) the length of Anne Boleyn's marriage to Henry VIII, from 23 January 1533 to 19 May 1536, when she was executed on a charge of adultery.

Thousand Island dressing a salad dressing with ketchup, chopped gherkins, and various other ingredients (probably named after the **Thousand Islands**).

Thousand Islands the group of about 1500 islands in the Upper St Lawrence River, on the border between the US and Canada.

thousand-legs a dialect term for a millipede.

Thousand-Year Reich a name for the **Third Reich**, which the Nazis envisaged as lasting for 1000 years.

War of a Thousand Days the Colombian Civil War of 1899 to 1903.

thousand and one very many, implying great variety, both 'quality and quantity'; found in some book titles, such as 'A Thousand and One Questions Answered on Astronomy' (compare **hundred and one**).

'Thousand and One Nights' the 'Arabian Nights' Entertainment'; originally simply a large number of stories, but later enlarged to make exactly 1001; the stories are told over 1001 nights by Scheherazade, the vizier's daughter, to avoid being killed by her husband, Schahirah, the sultan, who believes that no woman is virtuous and who marries a new wife every night and has her strangled at dawn; Scheherazade leaves each night's story incomplete, promising to complete it the following night; one story that she tells is that of **'Ali Baba and the Forty Thieves'**.

Two Thousand Guineas the annual horserace at Newmarket, first run in 1809 (with a prize of 2000 guineas or £2100) (compare **One Thousand Guineas**).

sixty-four thousand dollar question the most important question (the expression evolved from an uprated version of the US radio quiz programme which originally offered a top prize of $64 (see **sixty-four dollar question**); 'The $64,000 Question' was first broadcast in 1955).

QUOTATIONS AND SAYINGS

For one day in thy courts: is better than a thousand. (*Psalm* 84:10, Book of Common Prayer)

He carries weight! he rides a race!
'Tis for a thousand pound!
(William Cowper, *John Gilpin*, 1782)

Ring out the thousand wars of old,

Ring in the thousand years of peace.
(Tennyson, *In Memoriam*, 1850)

Where, where was Roderick then?
One blast upon his bugle-horn
Were worth a thousand men!
(Walter Scott, *The Lady of the Lake*, 1810)

A thousand shall fall at thy side, and ten thousand at thy right hand, but it shall not come nigh thee. (Bible: *Psalms* 91:7)

Saul hath slain his thousands, and David his ten thousands. (Bible: *1 Samuel* 18:7)

When I meet a man who makes a hundred thousand a year, I take off my hat to that man. (Bernard Shaw, *Heartbreak House*, 1919)

And the number of the army of the horsemen were two hundred thousand thousand: and I heard the number of them. (Bible: *Revelation* 9:16)

MILLIONS

EXPRESSIONS

feel like a million dollars to feel very well, be in excellent spirits.

Million Act the act of 1694 authorizing the holding of a lottery in this and subsequent years, by selling lottery tickets at £10 each, the prize being £1 million, a vast sum for those days.

million to one chance a highly unlikely chance.

thanks a million thank you very much (often used ironically).

QUOTATIONS AND SAYINGS

The first guinea is sometimes more difficult to acquire than the second million. (Rousseau, *Discours sur l'origine et les fondements de l'inégalité parmi les hommes*, 1755)

In all the world there is nothing more timorous than a million dollars, except ten million. (G W Johnson, *American Freedom and the Press*, 1958)

Fifty million Frenchmen can't be wrong. (saying popular among US soldiers in the First World War; the figure, which varied, was intended to represent the French population)

And they blessed Rebekah, and said unto her, Thou art our sister, be thou the mother of thousands of millions. (Bible: *Genesis* 24:60; this is the only instance of the word 'million' in the Bible)

9 READING AND VIEWING BY NUMBERS

NUMERICAL LITERARY AND ARTISTIC TITLES

The titles of many books, plays, films, radio and TV programmes, and musical compositions, to name but a few, frequently contain a number, often as a reference to the story or plot. Here are some of the best known, together with an explanation of the reference where helpful.

NOVELS AND STORIES (with year of publication)

1 *One Day in the Life of Ivan Denisovich*, Alexander Solzhenitsyn (1962)
One Dollar's Worth, O Henry (1910)

2 *A Tale of Two Cities*, Charles Dickens (London and Paris) (1859)
Two on a Tower, Thomas Hardy (1882)
Two Serious Ladies, Jane Bowles (1943)
Two Views, Uwe Johnson (man's and woman's viewpoints) (1965)
Two Years Before the Mast, Richard Henry Dana (1840)
Under Two Flags, Ouida (English and French) (1867)

3 *Soldiers Three*, Rudyard Kipling (1890)
The Three Clerks, Anthony Trollope (1857)
Three Lives, Gertrude Stein (1909)
Three Men in a Boat, Jerome K. Jerome (1889)
The Three Musketeers, Alexandre Dumas (1844)
Three Soldiers, John Dos Passos (1921)

4 *The Four Armourers*, Francis Beeding (1930)
The Four Feathers, A E W Mason (1902)
Four Frightened People, E. Arnot Robertson (1931)
The Four Just Men, Edgar Wallace (1905)
The Sign of Four, Conan Doyle (1890)

5 *Anna of the Five Towns*, Arnold Bennett (1902)
The Five Nations, Rudyard Kipling (1903)
The Five Orange Pips, Conan Doyle (1892)
The Five Red Herrings, Dorothy L. Sayers (1931)
A Five-year Sentence, Bernice Rubens (1978)

6 *A Set of Six*, Joseph Conrad (1908)
Sleeps Six, Frederic Raphael (1979)

7 *The House of the Seven Gables*, Nathaniel Hawthorne (1851)
Seven Days in New Crete, Robert Graves (1949)
Seven Men, Max Beerbohm (1919)
Seven Poor Travellers, Charles Dickens (1854)
The Seven Seas, Rudyard Kipling (1896)
The Seven Who Fled, Frederic Prokosch (1937)

8 *Eight Mortal Ladies Possessed*, Tennessee Williams (1974)
Two Tales and Eight Tomorrows, Harry Harrison (1965)

9 *Nine Coaches Waiting*, Mary Stewart (1958)
Nine Months in the Life of an Old Maid, Judith Rossner (1969)
The Nine Tailors, Dorothy L. Sayers (1934)

10 *Ten Little Niggers*, Agatha Christie (1939)
Ten Miles from Anywhere, P H Newby (1958)
Ten Minute Alibi, Anthony Armstrong (1933)
Ten North Frederick, John O'Hara (1955)

11 *Eleven*, Patricia Highsmith (1970)

12 *Twelve Horses and the Hangman's Noose*, Gladys Mitchell (1956)
Twelve Stories and a Dream, H G Wells (1901)

13 *No. 13*, M R James (a ghost story) (1910)
Thirteen Questions of Love, Boccaccio (*c.* 1336)

15 *Fifteen*, Beverly Cleary (1956)

17 *Seventeen*, Booth Tarkington (1916)

19 *1984*, George Orwell (1949)

20 *Twenty Years After*, Alexandre Dumas (after *The Three Musketeers*) (1845)

22 *Catch-22*, Joseph Heller (1961)

26 *Twenty-Six Men and a Girl*, Maxim Gorky (1899)

30 *30 Manhattan East*, Hillary Waugh (1968)

39 *The Thirty-Nine Steps*, John Buchan (1915)

45 *The Forty-Five*, Alexandre Dumas (1848)

80 *Around the World in Eighty Days*, Jules Verne (1873)

93 *Ninety-Three*, Victor Hugo (1793) (1879)

100 *The One Hundred Dollar Bill*, Booth Tarkington (1923)

101 *The One Hundred and One Dalmatians*, Dodie Smith (1956)
1000 *One Thousand Dollars*. O Henry (1908)
10,000 *Ten Thousand a Year*, Samuel Warren (1841)
20,000 *Twenty Thousand Leagues Under the Sea*, Jules Verne (1870)
4m *The Four Million*, O Henry (1906)
100m *A Hundred Million Francs*, Paul Berna (1955)
9000m *The Nine Billion Names of God*, Arthur C. Clarke (1955)

PLAYS

1 *Fanny's First Play*, Bernard Shaw (1912)
The First Man, Eugene O'Neill (1922)
2 *The Eagle Has Two Heads*, Jean Cocteau (1946)
A Memory of Two Mondays, Arthur Miller (1955)
The Second Mrs Tanqueray, Arthur W. Pinero (1893)
The Two Gentlemen of Verona, Shakespeare (*c.* 1592)
3 *Three Men on a Horse*, George Abbott (1935)
The Three Sisters, Anton Chekhov (1901)
4 *The Four Ps*, John Heywood (full title: *The Playe called the foure P.P.; a newe and a very mery enterlude of a palmer, a pardoner, a potycary, a pedler*) (1520)
The Fourth Wall, A A Milne (1928)
The Love of Four Colonels, Peter Ustinov (1951)
5 *The Fifth Column*, Ernest Hemingway (1940)
Five Finger Exercise, Peter Shaffer (1958)
6 *Six Characters in Search of an Author*, Luigi Pirandello (1921)
7 *Home at Seven*, R C Sherriff (1950)
Seven Against Thebes, Aeschylus (467 BC)
Seventh: Thou Shalt Steal a Little Less, Dario Fo (see Ten Commandments, p. 118) (1964)
8 *The Eighth of January*, R P Smith (1829)
10 *The Tenth Man*, Paddy Chayevsky (1959)
12 *The Twelfth Hour*, Alexei Arbuzov (1959)
Twelfth Night, Shakespeare (*c.* 1600)
13 *The Thirteenth Chair*, Will Irwin and Bayard Veiller (1916)
27 *Twenty-Seven Wagons Full of Cotton*, Tennessee Williams (1955)
38 *Amphitryon 38*, Jean Giraudoux (1929) (the play, about the Theban prince Amphitryon, had been dramatized 37 times before, including versions by Plautus, Molière and Dryden)
40 *Forty Years On*, Alan Bennett (1968)
77 *77 Park Lane*, Walter Hackett (1928)
1000 *Anne of the Thousand Days*, Maxwell Anderson (1948)
Marco Millions, Eugene O'Neill (1928)

FILMS

Films with numerical titles divide into two kinds. Those beginning with a number, and those that contain a number but not as the first word. As the first word of a film title is often the most significant, the number relates directly to the subject or story of the film, even where, by way of box office allurement, the numbered object is not stated. Of course, some movie titles are those of the novels or stories (or plays) on which they are based, and may be identical. Others are deliberately altered (for examples, see below, p. 141).

BEGINNING WITH A NUMBER (with year of first screening)

1 *One Flew Over the Cuckoo's Nest*, 1975 (suspect rapist is transferred to state mental hospital)
One is a Lonely Number, 1972
One, Two, Three, 1961 (Western executive, his boss's daughter, and Communist she wants to marry)
2 *Two and Two Make Six*, 1961 (two couples swap partners)
Two for the Road, 1966 (couples drive through France)
Two Seconds, 1932 (last two seconds of a criminal's life)
Two Way Stretch, 1960 (three convicts break jail; see p. 93)
3 *Three Coins in the Fountain*, 1954 (three girls find romance in Rome; the reference is to the Trevi Fountain there, whose own name means 'three'; a subtle title, therefore!)
The Three Faces of Eve, 1957 (three personalities of one woman)
Three into Two Won't Go, 1969 (husband has affair with girl, who moves into his and his wife's house)
Three Little Words, 1950 ('I love you'!)
3.10 to Yuma, 1957 (train departure time)
Trio, 1950 (three Somerset Maugham stories, after *Quartet*)
4 *Four Daughters*, 1938 (small-town family)
5 *Fifth Avenue Girl*, 1939 (unemployed girl poses as gold digger)

Five, 1951 (five survivors of nuclear holocaust)
Five and Ten, 1931 (chainstore)
Five Easy Pieces, 1970 (five romantic encounters)
6 *Six of a Kind*, 1934 (six people drive across USA)
7 *Seven Brides for Seven Brothers*, 1954
Seven Days in May, 1964 (political thriller)
Seven Days to Noon, 1950 (atomic bomb detonation threat)
Seven Samurai, 1954 (compare *The Magnificent Seven*)
The Seventh Cross, 1944 (seven Germans escape)
The Seventh Seal, 1957 (see Bible: *Revelation* 8:1)
The Seventh Veil, 1945 (compare the dance, p. 111)
8 *8½*, 1963 (number of films made previously by director Fellini)
Eight Iron Men, 1952 (eight US infantrymen wait for relief in Italian village)
Eight on the Lam, 1966 (bank teller brings up seven children)
9 *9½ Weeks*, 1986 (love affair)
Nine Men, 1943 (sergeant and eight men defend fort in Libyan desert)
Nine to Five, 1980 (office women plot to get rid of boss)
10 *'10'*, 1979 (composer rates girls from 1 to 10: finds '10')
Ten Rillington Place, 1970 (Christie murders address)
11 *Eleven Harrowhouse*, 1974 (jewel robbery address)
12 *Twelve Angry Men*, 1957 (about a jury)
Twelve O'Clock High, 1949 (about US bomber unit)
13 *The Thirteenth Letter*, 1951 (a poison pen letter)
14 *Fourteen Hours*, 1951 (man on window ledge)
29 *29 Acacia Avenue*, 1945 (domestic comedy)
30 *-30-*, 1959 (scoops at night in newsroom)
Thirty is a Dangerous Age, Cynthia, 1967 (timid nightclub pianist has 'woman trouble')
Thirty Seconds Over Tokyo, 1944 (air attack on Japan)
40 *Forty Carats*, 1973 (40-year-old woman has affair with 22-year-old man)
41 *The Forty-First*, 1927 (girl sniper gets 41st victim)
42 *Forty-Second Street*, 1933 (backstage musical)
49 *Forty-Ninth Parallel*, 1941 (stranded U-boat men cross border into USA)
50 *Fifty Roads to Town*, 1937 (mistaken identities in show-bound cabin)
55 *55 Days at Peking*, 1962 (Boxer siege in 1900)
60 *Sixty Glorious Years*, 1938 (on reign of Queen Victoria)
99 *99 River Street*, 1953 (diamond robbery address)
100 *One Hundred Men and a Girl*, 1937 (girl founds orchestra)
101 *One Hundred and One Dalmatians*, 1961 (London dogs save stolen Dalmatian puppies)
400 *The Four Hundred Blows*, 1958 (French film bases title on phrase *faire les quatre cents coups*, 'live in disorder')
2001 *2001: A Space Odyssey*, 1968 (year)
5000 *The Five Thousand Fingers of Doctor T*, 1953 (about 500 boys imprisoned in a castle of musical instruments)
1m *Le Million*, 1931 (about a final lottery ticket)
The Million Pound Note, 1954 (man inherits million dollars as a single banknote)
One Million BC, 1940 (remade as *One Million Years BC*)
50m *Fifty Million Frenchmen*, 1931

FILMS WITH NUMBER ELSEWHERE IN TITLE
(with screening year)

1 *Act One*, 1963 (writer grabs Broadway fame)
Public Hero Number One, 1935 (G-man tracks down gang)
2 *Abroad With Two Yanks*, 1944 (pun on 'a broad')
Between Two Worlds, 1944 (this world and the next)
Chapter Two, 1979 (a novelist's second affair)
Double Indemnity, 1944 (insurance fraud drama)
A Kid for Two Farthings, 1955 (see Bible: *Matthew* 10:29)
Only Two Can Play, 1962 (love affair)
Paradise for Two, 1937 (millionaire posing as reporter has to pose as a millionaire)
The Postman Always Rings Twice, 1946 (adultery and murder)
Tea for Two, 1950 (girl is promised money if she says 'no' to all questions for 24 hours)
You Only Live Twice, 1967 (James Bond adventure)

3 *And Baby Makes Three*, 1950 (pregnant wife divorces husband)
Close Encounters of the Third Kind, 1977 (direct contact between space aliens and humans on Earth)
Count Three and Pray, 1955 (Girl War soldier becomes parson)
The Devil Makes Three, 1952 (here: an obscure title)
Down Three Dark Streets, 1954 (three murders to be solved)
Fiddlers Three, 1944 (sequel to *Sailors Three*)
From Noon till Three, 1976 (murder and romance in 3-hour period)
A Letter to Three Wives, 1949 (three wives learn from friend that she has eloped with one of their husbands)
Paradise for Three, 1937 (romance in Germany)
Sailors Three, 1940 (drunken sailors capture German ship by mistake)
Sergeants Three, 1961 (cavalry sergeants beat hostile Indians)
Soldiers Three, 1951 (North-West Frontier years)
These Three, 1936 (schoolgirl and two teachers)
The Unholy Three, 1925 (ventriloquist, dwarf, strong man)

4 *Adam Had Four Sons*, 1941 (widower's family is cared for by governess)
Quartet, 1948 (four stories by Somerset Maugham)

5 *The Beast with Five Fingers*, 1946 (a severed hand murders)
Cleo from Five to Seven, 1961 (from 5.00 to 7.00)
Come Back to the Five and Dime, Jimmy Dean, Jimmy Dean, 1983
Count Five and Die, 1957 (British Intelligence uncovers double agent)
Pilot Number Five, 1943 (pilot volunteers for dangerous South Pacific mission)
Quintet, 1979 (about five people)
The Sheep Has Five Legs, 1954 (about quintuplets, but as French film, a punning title, for *mouton à cinq pattes* means 'unusual thing', 'rare bird')
Slaughterhouse Five, 1972 (anti-war fantasy)
Table for Five, 1983 (divorced husband and four children)
A Tale of Five Cities, 1951 (Rome, Vienna, Paris, Berlin, London)

6 *The Deep Six*, 1958 (about submarines)
The Inn of the Sixth Happiness, 1958 (Chinese rating)
The Secret Six, 1931 (about gangsters)
With Six You Get Egg Roll, 1968 (widow with three sons marries widower with one daughter)

7 *The Magnificent Seven*, 1960 (seven American bandits; based on *Seven Samurai*, see above)
Port of Seven Seas, 1938 (Marseilles)
Snow White and the Seven Dwarfs, 1937 (originally nameless, Walt Disney named the seven distinctively: Doc, Grumpy, Sneezy, Dopey, Bashful, Sleepy, Happy)

8 *Bluebeard's Eighth Wife*, 1938 (in original story, he murdered six wives before 7th escaped him)
Butterfield Eight, 1960 (society call girl's telephone number)
Dinner at Eight, 1933 (dinner party guests in dramatic situations)
Life Begins at Eight Thirty, 1942 (actor is forced to play Santa Claus on street daily at 8.30)
When Eight Bells Toll, 1971 (naval tale)

9 *At the Stroke of Nine*, 1957 (mad pianist plots murder of girl reporter)
The Man With Nine Lives, 1940 (like a cat)

10 *Force Ten from Navarone*, 1978 (on Beaufort Scale)
Slaughter on Tenth Avenue, 1957
Ten Little Niggers, 1945

11 *Ocean's Eleven*, 1960 (gang robbery)
On Friday at Eleven, 1960 (gang robbery)
Riot in Cell Block Eleven, 1954 (prison riot)

12 *Beneath the Twelve Mile Reef*, 1953 (romance and adventure for Florida sponge fishers)
Cheaper by the Dozen, 1950 (12 children in one family)
The Dirty Dozen, 1967 (12 convicts on suicide mission)

13 *Assault on Precinct 13*, 1976 (gang attacks police station)
Friday the Thirteenth, 1933 (characters in bus crash relive their lives)
Operator 13, 1934 (about woman spy)

15 *Angels One Five*, 1952 (about RAF in Second World War; air force jargon for 15,000 feet)

17 *Number Seventeen*, 1932 (girl jewel thief helps police)

Stalag 17, 1953 (US servicemen in Nazi POW camp)
Summer of the Seventeenth Doll, 1959 ('woman trouble')
21 *Over Twenty-One*, 1945 (an 'adult' comedy)
24 *Hill 24 Doesn't Answer*, 1954 (Israel fights the Arabs)
42 *Summer of '42*, 1971 (adolescents discover sex in 1942)
44 *Class of '44*, 1973 (sequel to *Summer of '42*)
45 *Colt 45*, 1950 (gun of the name had a calibre of .45)
73 *Winchester 73*, 1950 (western; name is that of rifle)
80 *Around the World in Eighty Minutes*, 1931 (compare book title, *Around the World in Eighty Days*, p. 136)
Star 80, 1983 (murder of model Dorothy Stretton in 1980)
99 *Convict 99*, 1938 (mistakenly appointed prison governor lets convicts take over)
99 and 44/100 Per Cent Dead, 1974 (gangster melodrama, with title spoofing famous ad.)
110 *Across 110th Street*, 1972 (police melodrama in New York)
123 *The Taking of Pelham 123*, 1974 (on New York subway train)
373 *Badge 373*, 1973 (on police detective)
451 *Fahrenheit 451*, 1966 (alleged temperature at which books burn, which fireman has to do in a future state)
880 *Mister 880*, 1950 (about a counterfeiter)
1000 *The House of a Thousand Candles*, 1936 (man has to live in unfurnished house to inherit from grandfather)
Man of a Thousand Faces, 1957 (Lon Chaney biopic)
2000 *Automania 2000*, 1963 (set in year 2000)
Death Race 2000, 1975 (set in year 2000)
7000 *Red Line 7000*, 1965 (about stock-car racer)
20,000 *The Beast from Twenty Thousand Fathoms*, 1953 (atom bomb test heat wakes prehistoric Arctic monster)
1m *The Beast with a Million Eyes*, 1955 (love defeats space monster)
The Boy Who Stole a Million, 1960 (set in Spain)
Brewster's Millions, 1935 (if playboy can spend a million pounds in two months, he will inherit many millions more)
If I Had a Million, 1932 (dollars)
6m *Symphony of Six Million*, 1932 (about New York)
1000m *Billion Dollar Brain*, 1967 (US megalomaniac plans world domination)
The Billion Dollar Scandal, 1932 (millionaire involves ex-convicts in shady oil deal)

NUMERICAL NONSENSE IN OPERA

Most operas have meaningful titles, whether containing a number or not. But the American composer, Virgil Thomson, wrote an opera that makes absolutely no sense, in either name or content.

It is called *Four Saints in Three Acts*, and has a libretto written by Gertrude Stein. Its characters include *five* saints (St Teresa I, St Teresa II, St Ignatius Loyola, St Chavez, St Settlement) and it is in *four* acts, not three. It is set in no particular time period and no particular place, although Spain has been suggested.

It was first performed in 1934 at Hartford, Connecticut, by the Society of Friends and Enemies of Modern Music.

FILMS GIVEN A NUMBER NAME BUT BASED ON A BOOK WITHOUT ONE

	film	book
1	*One Desire* (1955)	*Tracy Cromwell*
	One Eyed Jacks (1960)	*The Authentic Death of Hendry Jones*
	One More Tomorrow (1949)	*Animal Kingdom*
2	*Circle of Two* (1982)	*A Lesson in Love*
	Only Two Can Play (1961)	*That Uncertain Feeling*
	Trouble for Two (1936)	*The Suicide Club*
	The Two Faces of Dr Jekyll (1958)	*Dr Jekyll and Mr Hyde*
	Two Kinds of Women (1932)	*This is New York*
	Two Left Feet (1965)	*In My Solitude*
	Two-Letter Alibi (1962)	*Death and the Sky Above Us*
	Two Loves (1960)	*Spinster*
3	*The Devil Makes Three* (1952)	*Kiss of Death*
	Honeymoon for Three (1941)	*Goodbye Again*
	These Three (1936)	*The Children's Hour*

	Third Man on the Mountain (1959)	*Banner in the Sky*
	Three Coins in the Fountain (1954)	*Coins in the Fountain*
	Three in a Cellar (1970)	*The Late Boy Wonder*
	The Three Lives of Thomasina (1963)	*Thomasina*
	The Three Worlds of Gulliver (1959)	*Gulliver's Travels*
4	*Adam Had Four Sons* (1941)	*Nor Perfume Nor Wine*
	Four Hours to Kill (1935)	*Small Miracle*
5	*The Fifth Musketeer* (1978)	*The Man in the Iron Mask*
	Five Fingers (1952)	*Operation Cicero*
6	*The Inn of the Sixth Happiness* (1958)	*The Small Woman*
	Six Bridges to Cross (1954)	*Anatomy of a Crime*
7	*Gold of the Seven Saints* (1961)	*Desert Guns*
	Seven Alone (1974)	*On to Oregon*
	Seven Brides for Seven Brothers (1954)	*The Sobbin' Women*
	Seven Days' Leave (1930)	*The Old Lady Shows Her Medals*
	Seven Thieves (1960)	*Lions at the Kill*
	The 7th Dawn (1964)	*The Darian Tree*
	The Seventh Sin (1957)	*The Painted Veil*
8	*Eight Iron Men* (1952)	*A Sound of Hunting*
10	*Slaughter on 10th Avenue* (1957)	*The Man Who Rocked the Boat*
11	*On Friday at 11* (1961)	*The World in My Pocket*
13	*13 West Street* (1962)	*The Tiger Among Us*
20	*Love Begins at 20* (1936)	*Broken Dishes*
21	*Twenty-One Days* (1940)	*The First and the Last*
	21 Hours at Munich (1976)	*The Blood of Israel*
29	*29 Acacia Avenue* (1945)	*Acacia Avenue*
80	*The Lawless Eighties* (1957)	*Brother Van*
100	*100 Rifles* (1969)	*The Californio*
109	*PT 109* (1963)	*The Wartime Adventures of President John F. Kennedy*
999	*Dial 999* (1955)	*The Way Out*
1000	*I Died a Thousand Times* (1955)	*High Sierra*
20,000	*The Beast from 20,000 Fathoms* (1953)	*The Foghorn*
40,000	*Mayday: 40,000 Ft* (1976)	*Jet Stream*
80,000	*80,000 Suspects* (1963)	*The Pillars of Midnight*
1m	*Talk of a Million* (1951)	*Money Doesn't Matter*

BOOKS WITH NUMBER TITLES RENAMED AS FILMS WITHOUT

	book	**film**
2	*Second Man*	*He Knew Women* (1930)
	Two Hours to Doom	*Dr Strangelove* (1963)
3	*My Three Angels*	*We're No Angels* (1955)
	3rd Avenue, New York	*Easy Come, Easy Go* (1947)
	The Third Round	*Bulldog Drummond's Peril* (1938)
	Three Cups of Coffee	*A Woman's Angle* (1951)
	The Three Godfathers	*Hell's Heroes* (1930)

	The Three Oak Mystery	*Marriage of Convenience* (1960)
	Viper Three	*Twilight's Last Gleaming* (1977)
4	*4.50 From Paddington*	*Murder She Said* (1961)
	Four Punters Are Missing	*Who's Got the Action* (1963)
7	*Jewel of the Seven Stars*	*The Awakening* (1980)
	7½ Cents	*The Pajama Game* (1957)
	Seven Days to a Killing	*The Black Windmill* (1974)
	Seven Men at Daybreak	*Operation Daybreak* (1975)
	The Seven Pillars of Wisdom	*Lawrence of Arabia* (1964)
	We Are Seven	*She Didn't Say No!* (1958)
10	*Ten Against Caesar*	*Gun Fury* (1954)
	Ten Plus One	*Without Apparent Motive* (1972)
	Ten Second Jailbreak	*Breakout* (1975)
16	*Sixteen Hands*	*I'm From Missouri* (1939)
20	*Twenty Plus Two*	*It Started in Tokyo* (1961)
36	*North of 36*	*Conquering Horde* (1931)
50	*The Fifty-Minute Hour*	*Pressure Point* (1963)
1000	*A Thousand Shall Fall*	*Hangmen Also Die* (1943)
	Event 1000	*Gray Lady Down* (1977)

RADIO PROGRAMMES (broadcast by BBC in/from years stated)

1 *One Minute Please* (1951)
2 *Two-Way Family Favourites* (1960)
3 *Third Division* (1949)
7 *Monday Night At Seven* (1937–8)
8 *Monday Night At Eight* (1939–48)
20 *Twenty Questions* (1947–76)
30 *Thirty Minutes Worth* (with Harry Worth) (1963)
100 *Your Hundred Best Tunes* (1959)

TV PROGRAMMES (with years and explanation as necessary)

1 *First Love* (1982–) (four love stories)
One Man and His Dog (1982–) (BBC-produced sheepdog trials)
One Pair of Eyes (1967–75) (BBC-produced personal accounts)
2 *Never the Twain* (1981) (ITV comedy series)
Second City Firsts (BBC plays from Birmingham, Britain's 2nd city)
The Two Ronnies (1970s–80s) (BBC comedy series with Ronnie Barker, Ronnie Corbett)
Two's Company (1976–8) (ITV comedy: US authoress + UK butler)
3 *My Three Sons* (1960–71) (US)
Take Three Girls (1969–71) (BBC series)
Three After Six (1964–6) (three talkers after ITV 6.00 news)
Three Minute Culture (1989) (alleged frequency of channel-hopping by US viewer) (US)
Three of a Kind (1981) (three comedians in BBC series)
Three Piece Suite (1977) (three BBC comedy sketches)
3-2-1 (1978–) (ITV game show with 'countdown')
Three's Company (1977–) (young man and two girls in a flat) (US)
Triangle (1981–4) (ferry plies triangular schedule in BBC series)
5 *The Five Minute Show* (1989) (five-minute discussion on ITV)
Police 5 (mid-1960s–) (five-, then ten-minute ITV police report)
6 *Six English Towns* (1979–81) (BBC-produced architectural account of Chichester, Ludlow, Richmond (Yorkshire), Stamford, Tewkesbury, Totnes; a sequel was *Six More English Towns*)
Six Faces (1972) (BBC series with one man seen by six individuals)
Six-Five Special (1957–8) (BBC pop music show at 6.05 p.m.)
Take Six (1979–80) (six short story-films on ITV)

7 *Seven Days* (1982–) (C4 weekly religious programme)
Seven Faces of Jim (1962–3) (BBC series of seven comedy programmes starring Jimmy Edwards)
Seven Up (1963) (ITV documentary about seven different children; sequels at seven-year intervals were: *Seven Plus Seven*, *Twenty One*, *Twenty Eight Up*)
8 *Eight is Enough* (1977–80) (widower with eight children) (US)
9 *Chelsea at Nine* (1957–64) (ITV-produced Chelsea Palace variety at 9.00 p.m.)
Nine o'Clock News (1971–) (main BBC evening news)
Not the Nine o'Clock News (1979–82) (alternative comedy, on BBC2 at same time as news on BBC1)
10 *News at Ten* (main ITV evening news, at 10.00 p.m.)
15 *Fifteen to One* (1988) (C4 quiz show with 15 contestants, eliminated gradually down to one)
16 *Sweet Sixteen* (1983) (BBC sitcom: woman tycoon has affair with man 16 years her junior)
20 *Twentieth Century* (1957–70) (US current affairs series)
Twenty-Twenty Vision (1982–3) (C4 current affairs)
24 *24 Hours* (1965–mid 70s) (nightly BBC current-affairs report)
30 *Thirty-Minute Theatre* (1967–72) (BBC half-hour plays)
thirtysomething (1988) (seven friends in their 30s) (US)
40 *Forty Minutes* (1982–) (BBC series of 40-minute documentaries)
50 *Hawaii Five-0* (1968–79) (adventures of special investigation unit of Hawaii state government) (US)
66 *Route 66* (1960–3) (two young wanderers travel across US in search of adventure) (US)
77 *77 Sunset Strip* (1958–63) (private eyes' address) (US)
80 *A Kick Up the Eighties* (1984) (BBC alternative comedy for 1980s)
Life Begins at Eighty (1950–6) (panel game with 80-year-olds) (US)
100 *100 Great Paintings* (1980–2) (100 ten-minute BBC programmes, each on a different painting)
222 *Room 222* (1969–72) (US High-School comedy)
2000 *Citizen 2000* (1982–) (periodic ITV documentary about life of children born 1982 who reach majority in 2000)
1m *The Millionaire* (1954–9) (US comedy based on movie *If I Had a Million*: secretary to multi-millionaire gives away one million dollars of his money, tax free, to a different needy person each week).

MUSICAL TITLES

Numbers probably play a greater role in the titles of musical works than in any other artistic sphere. Not only are famous symphonies, for example, usually designated by a number rather than a name (Beethoven's Fifth Symphony, Mozart's 39th Symphony in E flat major), but many compositions are listed or catalogued numerically, from hymns to the special Köchel (K) numbers assigned to Mozart's works, or the *Bach Werke Verzeichnis* (BWV) to those of Bach. Thus the hymn 'Abide With Me' is No. 27 in *Hymns Ancient and Modern*, and will be familiar to many under that number, and the Mozart symphony just mentioned is K543 in the Köchel catalogue. Bach's *Christmas Oratorio*, similarly, is BWV 248, and those who sing and play it, or know it well, will be aware of this list number.

But numbers can also occur in the more orthodox names of operas just as they can in other literary works, even though they may be in a foreign language. Some are as follows:

2 *Les Deux Aveugles de Tolède* (The Two Blind Men of Toledo), Méhul, 1806
Les Deux Journées (The Two Days), Cherubini, 1800
I Due Foscari (The Two Foscari), Verdi, 1844
I Due Litiganti (The Two Litigants), Sarti, 1782
The Two Widows, Smetana, 1874
3 *L'Amore dei tre re* (The Love of the Three Kings), Montemezzi, 1913
Die Dreigroschenoper (The Threepenny Opera), Weill, 1928
Die Drei Pintos (The Three Pintos), Weber, 1821
The Love of the Three Oranges, Prokofiev, 1921
4 *I Quattro Rusteghi* (The Four Curmudgeons), Wolf-Ferrari, 1906.

10 NUMBER THAT NAME

NAMES BASED ON NUMBERS

Earlier chapters have shown how days of the week and months of the year have in some languages been given names based on numbers. Other more obvious naming systems have similarly drawn on number-words for some of their content, and this chapter considers a few of the best known.

PERSONAL NAMES

It makes logical sense to name a firstborn child 'First' or 'One', or its equivalent in the relevant language, and to name subsequent children in numerical progression, 'Second' ('Two'), 'Third' ('Three'), and so on.

The **Romans** used such numerical names for the praenomen or personal name of their children, among other types of name, so that boys could be called Primus, Secundus, Tertius, Quartus, Quintus, Sextus, and so on, and girls (although less frequently) similarly, as Prima, Secunda, Tertia, and so on.

Names of this type are fairly frequently found in classical literature, and there are even two in the Bible: St Paul's secretary, for example, was called Tertius (*Romans* 16:23), and a man named Quartus is mentioned in the next verse.

Some of the Latin names became popular subsequently, especially Septimus ('Seventh') and Decimus ('Tenth'), even though in more modern times there may have been no specific reference to the particular number. Similarly, Octavia ('Eighth') and Nona ('Ninth') acquired a certain popularity as girls' names, although, again, the numerical reference may not have been intentional. After all, many people give names simply because they like the sound of them!

A variant on Prima is Una (literally 'One') as a girl's name, while for boys Otto sometimes developed as a form of Octavius, especially as an Italian name.

The **Greeks** also sometimes used numerical names, so that a twin could be called Didymus (from Greek *didymos*, literally 'double'). Hence the additional name given to the disciple Thomas in the Bible (*John* 20:24), where the Greek name is simply a translation of his own name, which has the same meaning, deriving from Hebrew *te'om*, 'twin'.

Today, if a numerical name of this type is given meaningfully, its reference can be extended to denote not just the order of the child's birth, but the day or month when he or she was born. A boy named Quintin, for example, may have been born on the 5th of the month, or in May, while a girl named Primula may be named not after the flower but because she was born on the 1st of the month, or in January, or even on New Year's Day!

Some number names have become famous through **art and literature**. Raphael's famous painting *The Sistine Madonna*, for example, is so called because it was painted for the church of St Sixtus ('Sixth') in Piacenza (Sixtus himself appears in the picture, kneeling to the right of the Virgin). While in *The Faerie Queene*, Spenser deliberately named Una thus to typify the singleness of the true religion she represented (Protestantism), by contrast with the enchantress Duessa, who stood for the false religion (Roman Catholicism).

The poet Ralph Waldo Emerson, too, uses the name Una meaningfully in his poem addressed to the lady, with two lines running:

> But one I seek in foreign places,
> One face explore in foreign faces.

SURNAMES

Number names are also sometimes found as **surnames**, such as Quartermain ('Four Hands'), although it must be said that many are direct borrowings from numerical place-names, such as Twyford and Fouracre (see below).

Some English names appear to be number names but are actually not so in origin. Sixsmith, for example, probably means 'sickle smith', while Nineham means 'dweller by the enclosed land' (Old English *innam*).

Occasionally, one comes across a neat number as a surname, such as Eighteen (noted in Patrick Hanks and Flavia Hodges, *A Dictionary of Surnames*, 1988), and similar names are found in languages other than English, such as the French surnames Deux, Cinq, Six, Huit and Dix (recorded in Albert Dauzat, *Dictionnaire étymologique des noms et prénoms de France*, 1951). Names like these would probably have originated as nicknames. Unless, of course, they are corruptions of some other name.

But the two well-known Russian surnames Tretyakov and Shostakovich almost certainly derive from root words meaning, respectively, 'third' and 'sixth', probably referring to the third and sixth child to be born.

FICTIONAL NAMES

In the world of **fiction**, and especially science fiction, characters are sometimes given names that are actual numbers, rather than words based on numbers. This obviously applies when the word 'agent' or something similar accompanies the number, or is implied. We all know James Bond as '007', the number that meant 'licensed to kill', and in the American television series *Get Smart*, Maxwell Smart was 'Agent Number 86', while his female sidekick was the glamorous 'Agent 99'.

Robots, too, are likely to have numbers rather than human names, such as the two called C-3PO and R2-D2 in the *Star Wars* trilogy. G-8 was the hero of the US pulp magazine *G-8 and His Battle Aces*, published in the 1930s and 40s, while X-9 was the hero of a newspaper comic strip, *Secret Agent X-9*, also appearing from the 1930s. (In the 1960s, however, the hero was given a name instead of a number, and the strip title was changed to *Secret Agent Corrigan*.)

On British radio, PC 49 was the familiar policeman hero of an adventure series that ran for four years from 1949 (and who possibly took his number arbitrarily from this year). But the German spy named Funf who appeared in the popular radio show *ITMA* (It's That Man Again) during the Second World War was simply given an amusing name, not one that was intended to convey the German for 'five' (*fünf*).

'SEVEN' OR 'OATH'?

Because the Hebrew words for 'seven' (*šéva'*) and 'to swear' (*šavóa'*) are very similar, the name of the biblical town of Beersheba (modern Beer-Sheva) could mean either 'well of the seven' or 'well of the oath'. If the former, the name could relate to the covenant made between Abraham and Abimelech regarding a well, for the ownership of which Abraham gave Abimelech seven ewe lambs (*Genesis* 21:25-32). But if the latter, the name could relate to the exchange of oaths at a well made between Abimelech and Isaac (*Genesis* 26:23-33). Similarly, the biblical personal names Elizabeth, Bathsheba and Sheba could equally contain the word for 'seven' or 'oath'. If 'seven', then this would be in the sense 'fullness', 'perfection', and Elizabeth would mean 'God is perfection', Bathsheba would mean 'daughter of perfection', and Sheba, which is both personal name and place-name, would mean simply 'perfection'.

PLACE-NAMES

The use of numbers to form a **place-name** is commoner than might be supposed, and many places have names that refer, for example, to a single farm, a double ford, a group of three bridges, a set of four stones, and so on. The actual number may have been distorted from its original form, and so not be recognizable, and such distortion is frequent in British names, for instance, that derive from Old English (Anglo-Saxon) or Scandinavian.

Here is a selection of British number-names, with the modern meaning given where it is not immediately obvious:

1 Aintree ('one tree'); Ancoats ('lonely huts'); Olney ('one glade'); Onehouse.

2 Llandeusant ('church of the two saints'); Penrhyndeudraeth ('promontory with two beaches'); Tiverton ('farm at the double ford'); Twemlow ('place by two hills'); Two Bridges; Two Dales; Two Gates; Two Tree Island; Two Waters; Twycross ('double cross'); Twyford ('double ford').

3 Llantrisant ('church of the three saints'); Three Bridges; Three Cocks; Three Holes; Three Legged Cross; Three Locks; Three Mile Cross; Threemilestones; Three Rivers.

4 Fardle ('fourth part'); Featherstone ('four stones'); Four Marks; Four Oaks; Fourstones; Four Wants ('four ways').

5 Fifehead ('five hides'); Fifield ('five hides'); Filey ('five clearings'); Fitzhead ('five hides'); Five Ashes; Five Bridges; Fivehead ('five hides'); Five Houses; Five Lanes; Five Oaks;

Five Roads; Five Ways; Fyfield ('five hides'); Plynlimmon ('five beacons').

6 Six Bells; Sixhills; Six Mile Bottom.

7 Seaborough ('seven hills'); Seavington ('settlement of the dwellers at the seven wells'); Sevenash; Seven Dials ('seven aspects'); Sevenhampton ('settlement of the dwellers at the seven wells'); Sevenoaks; Sevington ('settlement of the dwellers at the seven wells'); Sewell ('seven wells').

8 Eight Ash Green.

9 Nine Ashes; Ninebanks; Nine Elms; Nine Ladies; Nine Mile Bar; Noonstones ('nine stones'); Nynehead ('nine hides').

10 Stokeinteignhead ('farm of the ten hides'); Tenacre; Ten Mile Bank; Tinhead ('ten hides').

12 Twelveheads ('12 hammer-heads'); Twelve Oaks.

16 Sixteen Foot Drain.

20 Twenty; Twenty Foot River.

30 Piddletrenthide ('place of 30 hides by the river Piddle').

40 Forty Foot Drain.

48 Eight and Forty ('48 houses').

100 Hundred Acres; Hundred Foot Drain; Hundred House.

Like Britain, Ireland has her share of number-names, too. They include: Twomileborris, Three Castles, Fivemiletown, Sixmilebridge and Nine Mile House. In Connemara, a group of mountains is known as the Twelve Bens (or Twelve Pins).

Occasionally, one finds an apparent number-name that did not actually derive from a number. One example is the district of Ilford (London) called Seven Kings. The name is popularly associated with seven Saxon kings, but in origin means 'place of Seofeca's people'.

A sea area to the east of Scotland, with a name familiar from shipping forecasts, is Forties. The area's full name is Long Forties, and denotes a lengthy depression in the sea bed at a depth of around 40 fathoms.

Outside Britain, English-language number-names are usually more straightforward, if only because most of them have been given fairly recently, so have not been corrupted. Among them are the following (with country):

1 One Tree Hill (New Zealand).

2 Twin Falls (USA); Twin Lakes (USA); Twin Mountains (USA); Twofold Bay (Australia); Two Harbors (USA); Two Mountains (Canada); Two Rivers (USA).

3 Three Brothers (USA); Three Fingers (USA); Three Forks (USA); Three Kings Islands (New Zealand); Three Mile Island (USA); Three Rivers (Canada); Three Sisters (USA).

4 Four Corners (USA, at junction of four states: Colorado, New Mexico, Arizona, Utah); Four Lakes (USA); Four Peaks (USA).

5 Five Forks (USA).

7 Seven Devils Mountains (USA); Seven Hills (USA); Seven Isles (Canada); Seven Mile Beach (USA).

9 Nine Point Mesa (USA).

12 Twelve Apostles (USA: group of about 20 islands in Lake Superior).

24 Twenty-Four Parganas (India: *pargana* is a revenue district).

40 Forty Fort (USA: stockade was built by first 40 settlers).

49 Forty-Nine Creek (USA: wagon-train route of 1849).

George R. Stewart's *American Place-Names* mentions US names based on even higher numbers, such as Old Hundred, Hundred and Ten Mile Creek, and One Hundred and Forty Mile Creek. Professor Stewart regards the record, however, as Hundred Thousand Creek, in Alaska, although the name itself is of unknown origin.

Similar high-value names are held by the Thousand Islands, a group of around 1500 islands in the upper St Lawrence River, midway between the USA and Canada, while Thousand Oaks is a city in California and Thousand Ships Bay a bay in the Solomon Islands.

Number-names are also found in languages other than English. Here is a selection, with their country of location:

2 Bahrain (in Persian Gulf, 'two seas', i.e. either side); Biarritz (France, 'place of two rocks'); Doab (India and Pakistan, 'two waters', i.e. the Ganges and Yamuna Rivers); Ekibastuz (USSR, 'two-headed lake'); K2 (Pakistan, second mountain in Karakoram range to be surveyed, and also second highest there); Tuapse (USSR, 'two rivers'); Zweibrücken (W. Germany, 'two bridges').

3 Trento (Italy, 'three teeth', i.e. three-headed mountain); Triglav (Yugoslavia, 'three heads', i.e. three-headed mountain); Trinidad (Trinidad & Tobago, 'trinity', i.e. three mountain peaks on island); Tripoli (Libya, 'three towns', i.e. historic cities of Oea, Leptis Magna, and Sabrata); Tripoli (Lebanon, 'three towns', i.e. historic cities of Tyre, Sidon and Aradus); Tripolis (Greece, 'three towns'); Tripura (India, 'three towns').

4 Chardzhou (USSR, 'four streams'); Périgueux (France, 'four armies'; as name of the Petrocorii, a Gallic race); Pithiviers (France, 'fourth', from Gallic personal name: Petuario, 'fourth-born'); Shikoku (Japan, 'four provinces'); Vierwaldstätter See (Switzerland, 'lake of the four forest cantons', otherwise Lake Lucerne, which is enclosed by the cantons of Schwyz, Uri, Unterwalden, and Lucerne).

5 Fünfkirchen (Hungary, 'five churches', as former German name of Pecs); Pentapolis (ancient world, 'five towns', used in Italy for group of cities Rimini, Ancona, Fano, Pesaro, Senigallia; in Asia Minor for Cnidus, Cos, Lindus, Camirus, Ialysus; in Cyrenaica for Apollonia, Arsinoë, Berenice, Cyrene, Ptolemaïs); Punjab (India, 'five waters', for tributaries of Indus: Jhelum, Chenab, Ravi, Beas, Sutlej); Pyatigorsk (USSR, 'five mountains', i.e. one with five peaks).

6 Montevideo (Uruguay, said by some to derive from Spanish legend on map, *Monte VI de O*, '6th mountain from the west', referring to raised land at entrance to harbour).

7 Semipalatinsk (USSR, 'seven palaces'); Les Sept-Îles (France, 'the seven islands'); Siebenbürgen (Romania, 'seven fortified towns', otherwise Transylvania); Siebengebirge (W. Germany, 'seven mountain chains').

8 Tuvalu (Pacific Islands group, former Ellice Islands, 'eight standing together', referring to atolls here, in fact nine in number).

9 Kyushu (Japan, 'nine provinces').

10 Decapolis (ancient Palestine, 'ten towns', for league of ten Greek cities here formed after Pompey's campaign of 1st century BC).

12 Dodecanese (Greece, 'twelve islands', for main islands in group in Aegean Sea: Astypalaia, Kalymnos, Karpathos, Kasos, Khalke, Kos, Leros, Lipsos, Nisyros, Patmos, Syme, Telos).

33 Treinta y Tres (Uruguay, '33', referring to 33 independence fighters of 19th century).

Three number-names in the thousands are those of the Madagascan capital, Antananarivo, 'town of 1000' (i.e. inhabitants), Tyumen, in the USSR, '10,000', with reference to a warrior race of this number, and the Lakshadweep Islands (former Laccadives), in the Arabian Sea, where they belong to India, '100,000 islands', referring to their great number. (There are actually only 20 islands, but the name could originally have included the nearby Maldives, which comprise 19 clusters of coral atolls.)

NAMING THE TWINS

Among the Kachin people of Upper Burma, no child is ever given the same name as another. But difficulties arise when twins are born, since according to Kachin beliefs, twins are regarded as one human being in two persons. Special naming rites are brought into effect for twins, therefore. When the children reach the age of ten, they are invited three times to plunge their hands in turn into a jar containing a few grains of rice. One of the grains is painted red. Whichever of the twins brings it out first is the one to keep the name given at birth. The other twin from then on will be called simply 'Second'.

STREET NAMES

Many towns and cities of the world have streets with number-names such as First Street, Second Avenue, and so on. But probably nowhere more so than the United States, where streets such as New York's Fifth Avenue and 42nd Street are internationally famous. (Many such streets have their own entries in Chapter 8.)

How did the American fashion for street number-names start? They mostly owe their origin to William Penn, the founder (and namer) of Pennsylvania, who came to America from England in 1682 to establish his colony on land granted him by Charles II in lieu of a large debt repayment owed to his father.

Penn founded (and also named) Philadelphia, and when he arrived there he discovered that building was already in progress and that many of the streets had been named after important residents. William Penn, however, was a Quaker, and did not look favourably on a naming system that elevated some people above others like this. Instead, he proposed that the streets should be named numerically, especially as their symmetrical arrangement lent itself to such an orderly scheme. Beginning at the eastern boundary of the town, therefore, Penn named First Street, and named parallel streets in numerical sequence as he moved westward. For the north-to-south streets, however, he chose contrasting non-numerical names (mainly after trees and plants), in order to avoid confusion.

Philadelphia, whose own name means 'brotherly love', and similarly symbolizes Penn's Quaker beliefs, thus set the 'number-name' pattern for other cities built subsequently, so that numerical names are now found in many American cities.

In many cases, as in Philadelphia, the streets running at right-angles to the numbered streets usually have non-numerical names, even just a letter of the alphabet, as in Washington, DC. But in New York, there are number-names for both east–west and north–south streets, and the contrast is expressed by calling the north–south streets Avenues. Thus, on Manhattan Island, streets named First to Tenth Avenue run from north to south, with the numbering starting at the eastern boundary, as for Philadelphia. The east-west streets, however, remain Streets, with the numbering starting at the southern end of the Island and extending up over 200 at the northern end.

According to *The Street Directory of the Principal Cities of the United States* (1908), New York has more streets with number-names than any other city. It does not follow, however, that *all* American cities have street-names of this type, and Boston, Massachusetts, for example, has conventional names, after people and places, as in most British cities.

London, on the other hand, is large and varied enough to have a fair proportion of streets with number-names, and the index to the *Greater London Street Atlas*, published by the Automobile Association in 1983 (4th revised edition) shows that Greater London has:

25 First Avenues
19 Second Avenues
12 Third Avenues
9 Fourth Avenues
8 Fifth Avenues
5 Sixth Avenues
2 Seventh Avenues
2 Eighth Avenues
1 Ninth Avenue
0 Tenth Avenues

as well as very small representation of number-names with a word other than 'Avenue'.

Other cities have a smaller proportion of such names, although, again, usually with 'Avenue'. Thus Leeds has 2 First Avenues, 2 Second Avenues, 2 Third Avenues, 1 Fourth Avenue, 2 Fifth Avenues, 1 Sixth Avenue, 2 Seventh Avenues, 2 Eighth Avenues, 1 Ninth Avenue, 1 Tenth Avenue, 1 Eleventh Avenue, 1 Twelfth Avenue, 1 Thirteenth Avenue, 1 Fourteenth Avenue, 1 Fifteenth Avenue, 1 Sixteenth Avenue, 1 Seventeenth Avenue and 1 Eighteenth Avenue. Most of these are in a compact area in the district of Armley, west of the city centre.

Some new towns are allocated street number-names, too, such as Milton Keynes, which was planned on a grid layout, like many American cities. It has a run of parallel streets named, for example North Sixth Street, South Sixth Street, extending eastwards to North Thirteenth Street.

Some of the shorter London streets with number-names are actually named after inns or pubs. Examples are Three Colts Lane, Three Cups Yard, Three Kings Yard, Three Oak Lane and Three Tuns Court.

PUB NAMES

As mentioned above, many inns or pubs have number names, which are sometimes passed on to the street on which the inn is located.

Some of the pub names of this type simply took the number from that of the pub's address, as a house number. Others have names that refer to history or folklore. The *Dictionary of Pub Names* by Leslie Dunkling and Gordon Wright includes the following pub number-names among its hundreds of entries:

1 One and All, One and Only, One Bull, One Elm, One Tun.
2 Twa Corbies, Twa Dogs, Two Bells, Two Blues, Two Brewers, Two Chairmen, Two Friends, Two Mile Oak, Two Palfreys, Two Poplars, Two Puddings, Two Sawyers, Two Saxons, Two Ships, Two Woodcocks.
3 Three Arrows, Three Bells, Three Blackbirds, Three Bridges, Three Bucks, Three Castles, Three Chimneys, Three Cocks, Three Colts, Three Compasses (etc., over 50 altogether, as easily the most popular number).
4 Four Alls, Four Bells, Four Counties, Four Crosses, Four Elms, Four in Hand, Four Lords, Four Seasons, Four Sisters, Four Winds.
5 Five Alls, Five Arrows, Five Bells, Five Elms, Five Lions, Five Miles From Anywhere No Hurry, Five Pilchards, Five Ringers, Five Wand Mill.
6 Six Bells, Six in Hand, Six Lords, Six Packs, Six Ringers.
7 Seven Saxons, Seven Seas, Seven Sisters, Seven Stones, Seven Ways, Seven Wives.
8 Eight Bells, Eight Kings, Eight Locks, Eight Ringers.
9 Nine Elms, Nine Pins, Nine Saxons.
10 Tenbell.
11 Eleven Cricketers, Eleven Ways.
12 Twelfth Man, Twelve Bells, Twelve Knights.
13 Thirteenth Volunteer Mounted Cheshire Rifleman.

Higher numbers included pubs named Fifteen Balls, Sixteen String Jack, Nineteenth Hole, Twenty Churchwardens, Forty Foot, and Hundred House.

Like all pub names, most of these number-names lend themselves to a suitable pictorial representation on an inn sign. The many 'bell' names refer mostly to church bells and hand bells, rather than ship's bells, even though time is an important element in a pub's opening (and closing) hours!

WHICH SIXTUS WAS SIXTH?

There were five popes named Sixtus ('Sixth'). But which one was genuinely sixth?

The first three were saints, with Sixtus I living in the 2nd century, Sixtus II in the 3rd, and Sixtus III in the 5th. Sixtus IV lived in the 15th century, and Sixtus V in the 16th.

It seems likely that Sixtus I took his name as he regarded himself as the sixth pope after St Peter, who was the first. The other popes then took their names after him.

It was Sixtus II who gave his name to the church for which Raphael painted the *Sistine Madonna*. But it was Sixtus IV who gave his name to the Sistine Chapel, the pope's chapel in the Vatican that was so splendidly decorated by Michelangelo.

These five popes are the only ones to have had number names.

SHIP NAMES

Several ships have been given number-names, over the years, usually with a historical, literary or legendary reference, rather similar to inn and pub names. Among such names noted in Don H. Kennedy's *Ship Names*, published for The Mariners Museum, Newport News, Virginia, USA in 1974, are the following:

1 First Effort, First Whelp
2 Two Brothers, Two Marys, Twin Sons
3 Three Josephs, Three Ostrich Feathers, Three Sisters
4 Four Cousins, Four Sons, Fourth of July
5 Five Daughters, Five Kinds
7 Seven Generals, Seven Kings, Seven Sisters
9 Ninth Whelp
10 Ten Sisters, Tenth Whelp
12 Twelve Apostles
14 14 Ramadan
50 50 Let

The 'Whelps' were English coastguard ships built to one design in 1628 with the aim of suppressing French and Turkish privateers in home waters. The last-named ship in the list above, the 50 Let, was a Soviet container ship. Her name means '50 years', and marked the 50th anniversary of the Russian Revolution, i.e. the year 1967.

Some ships have purely numerical designations, however, such as the five American gunboats captured off New Orleans by the British in 1814. These were simply called No 5, No 23, No 156, No 162 and No 163. The US Navy also used names of this kind in the First World War, when over 100 patrol boats were named merely Eagle 1, Eagle 2, Eagle 3, and so on.

One bunkering barge operating out of San Pedro, California, and owned by the Union Oil Company, was named Nineteen Twenty-Three, the year she was built.

REGIMENTAL NAMES

In British military history, army regiments were originally named after their colonel. A ruling of 1751, however, directed that all regiments should be known by their numbers, or ranking, in the line, not by their colonels' names, while certain regiments could retain their former titular distinctions, such as '3rd Regiment of Foot'. But in 1881 numerical titles were abolished, and county names were introduced instead. Even so, a few regiments still cling (unofficially) to their old numerical titles today.

The following are the first 12 such numerical titles still found in use, together with their official titles, as listed in JMBrereton's *A Guide to the Regiments and Corps of the British Army on the Regular Establishment* (1985):

1st The Royal Scots (The Royal Regiment)
2nd The Queen's Royal Regiment (West Surrey)
3rd The Buffs (Royal East Kent Regiment)
4th The King's Own Royal Regiment (Lancaster)
5th The Royal Northumberland Fusiliers
6th The Royal Warwickshire Fusiliers
7th The Royal Fusiliers (City of London Regiment)
8th The King's Regiment (Liverpool)
9th The Royal Norfolk Regiment
10th The Royal Lincolnshire Regiment
11th The Devonshire Regiment
12th The Suffolk Regiment

As an example of the historic naming process, the following (with year of naming) is the record of the Royal Norfolk Regiment listed above, beginning with its colonel's name:

Colonel Henry Cornwall's Regiment of Foot (1685)
9th Regiment of Foot (1851)
9th (or East Norfolk) Regiment of Foot (1782)
The Norfolk Regiment (1881)
The Royal Norfolk Regiment (1935)

In 1959 the regiment amalgamated with The Suffolk Regiment to form the 1st East Anglian Regiment (Royal Norfolk and Suffolk), the Suffolk Regiment itself having earlier been the 12th Regiment of Foot in a similar manner. The Royal Norfolk and Suffolk Regiment then in turn amalgamated with four other regiments (in 1964) to become The Royal Anglian Regiment, which today unofficially retains the numbers of all previous individual regiments, as the 9th, 10th, 12th, 16th, 17th, 44th, 48th, 56th and 58th!

However, the above ruling applied only to infantry regiments (hence 'Regiment of Foot'), and other regiments still retain their numerical titles officially, such as the 1st, 2nd, 3rd, etc. Dragoon guards and other cavalry regiments. The 1st The Queen's Dragoon Guards, for example, was formed in 1959 by amalgamation of the 1st King's Dragoon Guards with The Queen's Bays (2nd Dragoon Guards).

THE ORIGINAL SEVEN SISTERS

There are at least three places in Britain called by the name of Seven Sisters. The first, and best known, is the range of chalk cliffs in Sussex (see p. 112). The second Seven Sisters is a district of north London, in the borough of Haringey. It takes its name from seven elm trees that once stood at the corner of Seven Sisters Road, near Page Green.

The third Seven Sisters is a village near Neath, in South Wales. It takes its name from seven real sisters, the daughters of David Evans, a coalmine owner, in the second half of the 19th century, when the mine was opened. They were Nancy Isabella, Mary Diana, Sarah Jane, Margreta, Frances Matilda, Maria Louisa and Sophia Annie. The last named lady was the one who lived longest, until 1947, and was the only one of the Seven Sisters who never married.

POP AND ROCK GROUPS

Regimental number-names (above) are rather esoteric. Much more widely known are the names of various pop and rock groups, many of which have incorporated a number. The number usually (but not always) refers to the number of members in the group. Where it does not, the reference is sometimes obscure or covertly sexual. Among the best-known number-names of pop groups have been the following, with explanatory notes in some cases:

2 U2 (meaning 'you too', as all fans could participate); Thompson Twins (four members took name from Tintin cartoon strip characters).

3 Big Three; Fun Boy Three; Three Degrees; Three Dog Night.
4 Four Preps; Four Seasons; Four Tops; Gang of Four.
5 Dave Clark Five; Fifth Dimension; Five Satins; Jackson Five; Jive Five; MC5 (i.e. 'Motor City Five', after their home base, Detroit); The Pentangle (five members); We Five.
7 Temperance Seven (tongue-in-cheek name for the nine-man group).
10 Ten Years After; 10cc.
17 Heaven 17.
40 UB40 (from number of UK unemployment benefit card).
42 Level 42.
52 B-52s (from the 1960s slang term for the lofty beehive hairstyle favoured by many girls, itself named after the huge American bomber; the group was not formed until 1976, however).
69 Sham 69 (from Hersham, Surrey; a punk group and name).
100 Haircut 100 (deliberately whimsical name).

BRAND NAMES

Several well-known brand names are based on numbers, and can have a wide range of references, from a historical dating to a manufacturing quantity (number of ingredients, products or varieties). The following are a selection, with the nature of the product and a brief account of the numerical reference, where not self-evident:

1 Pimm's No 1; Primus.
2 Double Century (sherry: blended in 1930 to mark 200th anniversary of founding of firm); Double Diamond (beer: conventional symbol, something like 'XXX' for a strong beer); Dubble Bubble (bubble gum: twice the strength of ordinary gum).
3 Mitsubishi (car: Japanese name means 'three diamonds' hence car's logo of three rhombuses); Sanyo (electrical goods: Japanese name means 'three oceans', denoting three main oceans of the world, Atlantic, Pacific, Indian, and so expressing company's international market); Three Candlesticks (stationery: from inscription on token coin, 'At the 3 Candlesticks', found at site of firm founder's house, London, in 1799); Three Castles (tobacco: from line in Thackeray's novel *The Virginians*: 'There's no sweeter tobacco comes from Virginia, and no better brand than the Three Castles'); Three Fives (cigarettes: based on engine number 999 of State Empire Express train on which firm founder used to travel); Three-in-One (oil: lubricates, cleans, preserves); 3M (adhesive tapes: abbreviation of *M*innesota *M*ining and *M*anufacturing Company); Three Nuns (tobacco: origin obscure); Triplex (safety glass: both in three layers and triple strong, like 'XXX' beer).
4 Tetrapak (carton: tetrahedral, or four-sided, in shape).
5 Chanel No 5; Pentax (camera: from its *penta*prism, or five-sided prism, which facilitates reflex viewing).
6 Subaru (car: from Japanese name for Pleiades, whose six main stars represent six companies which merged to form original firm in 1953, and which are seen in car's logo).
7 7-Up (soft drink: name said to have been invented after six earlier attempts at a name but undoubtedly linked with the card game; see p. 112).
8 Q8 (petrol: imported from Kuwait!).

Three names based on higher numbers have their stories, too.

Heinz 57 Varieties has an arbitrary number, devised by the firm's founder, Henry Heinz, when riding one day in 1897 in an elevated train in New York. He saw an advertisement for shoes in '21 Styles', and based his higher number on that, although knowing that his new firm already manufactured more than this number of varieties.

Vat 69 whisky was probably simply named after the particular vat that contained the chosen blend. Even so, the number, for obvious reasons, helps to sell the product.

4711 eau-de-Cologne got its number from the 'address' chalked on the wall of the original factory in Cologne by French soldiers who had difficulty reading the German script. The number was also said to be that of a secret recipe for making eau-de-Cologne. But, if so, it is doubtful that the French soldiers were aware of this.

NUMBERS AS AFFIXES

People with identical names are traditionally given a sequential (ordinal) number for purposes of differentiation, especially when the need to differentiate is important for some reason.

NUMBER NAMES IN THE NEWS

In the spring of 1989, *The Times* newspaper conducted a lively correspondence among its readers on people round the world with number names.

The paper's issue for 21 April contained no less than three letters on the subject. The first was from a Cornish reader whose Japanese grandmother was named Fumiko ('two-three child', as her mother was 23 when she was born) and whose sister was named Miyoko ('three-four child', as her mother was 34). The second letter, from a Berkshire reader, pointed out that the three youngest sons of Pascoe and Sofia Grenfell, who lived at Wilton Park, Beaconsfield, at the end of the 19th century, were named Robert Septimus ('7th'), Francis Octavius ('8th') and Riversdale Nonus ('9th'). These last two were twins.

The third letter, from a Surrey reader, told how some 45 years earlier a boy named Museveni was born in the Ankole district of Uganda, Africa. His name means 'soldier of the 7th', referring to the 7th Battalion of the King's African Rifles who were recruiting locally. He is now Lieutenant-General Yoweri Museveni, President and Minister of Defence of the Republic of Uganda.

Thus, in a single family, a name can be handed down from father to son, with the successive sons becoming, for instance, John H. Smith II, John H. Smith III, and so on. A differentiation between simply two can be made effectively by 'Snr' (Senior) and 'Jnr' (Junior), especially between father and son, just as some British public and preparatory schools use (or used) 'Ma.' (Major) and 'Mi.' (Minor) for the elder and younger of two brothers, respectively. But for three or more, an added number is the best and most common solution. In such instances, the number itself is conventionally written in Roman figures.

Where the person is a public figure, the need to differentiate is essential. Hence the ordinal suffixes used by **monarchs** and **popes**. Many such names have gone down in history, and most Britons (and doubtless several others) can distinguish, up to a point, between James I and James II, George III and George IV, Edward VII and Edward VIII. If only from schooldays or romantic fiction, Charles I is remembered as the king who was executed as a tyrant, as 'King Charles the Martyr', while Charles II has made his name both for his many mistresses and as (or hence) the 'Merry Monarch'. If, as is expected, the present heir to the throne succeeds to this position in due course, as the eldest son of Queen Elizabeth II, he will become Charles III.

Some monarchies have worked their way through an impressive number of identically named kings, with France, for example, having not only a Louis XVIII (died 1824) but a 'Louis XIX' (Duc d'Angoulême), the last dauphin of France, who renounced the throne in 1830 but who was accorded his title by Legitimists six years later.

The highest regnal numbers scored by other European states are (where at least III):

Belgium	Leopold III
Denmark	Christian X
England	Henry VIII
Germany	Frederick III
Great Britain	Edward VIII
Italy	Charles Emmanuel IV
Netherlands	William IV
Norway	Haakon VII
Portugal	John VI
Russia	Ivan VI
Scotland	James VI
Spain	Alfonso XIII
Sweden	Charles XVI Gustaf

There have been some similarly high-numbered popes, with John XXIII (died 1963) as the holder of the record. (His nominal predecessor, John XXII, lived some six centuries before him, dying in 1334.)

The overall record regnal number, however, belongs to Count Heinrich LXXV of Reuss, who ruled in Germany for the two years ending 1801. His principality was one of two of the name in what is now Thuringia, and the rulers of both princely houses were always called Heinrich (Henry), in honour of the Holy Roman Emperor Henry VI, who reigned in the 12th century. They were thus distinguished solely by regnal number.

BIBLIOGRAPHY

As there is scarcely an area of human knowledge and endeavour that does not involve numbers—geography, history, language, literature and the sciences, especially mathematics, are the ones that spring to mind—any bibliography on the subject must necessarily be selective.

The following titles, however, treat the field more obviously than many, and I have selected books that on the whole are approachable, interestingly written, and not too technical. I have marked with an asterisk (*) those titles that are particularly relevant or comprehensive.

After the listing, I give further suggestions for useful sources of information on numbers.

Achelis, Elisabeth, *Of Time and the Calendar*, Spearman, UK, 1955

Augarde, Tony, *The Oxford Guide to Word Games*, Oxford University Press, UK, 1984

Beiler, A. H., *Recreations in the Theory of Numbers*, Dover, USA, 1964

Bickerman, E. J., *Chronology of the Ancient World*, Thames & Hudson, UK, 1968

Bodmer, Frederick, *The Loom of Language*, Allen & Unwin, UK, 1944

Butler, Christopher, *Number Symbolism*, Routledge & Kegan Paul, UK, 1970

Cheiro's Book of Numbers, Barrie & Jenkins, UK, 1978

Churchill, Eileen Minna, *Counting and Measuring*, Routledge & Kegan Paul, UK, 1961

Cruden, Alexander, *Complete Concordance to the Old and New Testaments*, Lutterworth Press, UK, 1930

de Morgan, Augustus (comp.), *Book of Almanacs, with an Index of Reference*, Macmillan, UK, 1907

Devi, Shakuntala, *Figuring: The Joy of Numbers*, Deutsch, UK, 1977

Diringer, David, *Writing: A History of Writing*, Thames & Hudson, UK, 1962

Diringer, David and Freeman, H., *The History of the Alphabet*, Gresham Books, UK, 1977

Edwards, Willard E., *The Edwards Perpetual Calendar: A New Solution to an Old Problem, that of a Simplified, Practical, Perpetual Calendar, Easy to Use and Easy to Remember*, Willard Edwards, UK, 1943

Flegg, Graham, *Numbers and Counting*, Open University, UK, 1974

*Flegg, Graham, *Numbers: Their History and Meaning*, Penguin, UK, 1984

Fomin, S. V., *Number Systems*, University of Chicago Press, USA, 1975 (translated from the Russian)

Gaines, Helen Fourche, *Cryptanalysis: A Study of Ciphers and Their Solution*, Dover, USA, 1956

Gardner, Martin, *Mathematical Carnival*, Penguin, UK, 1978

Gardner, Martin, *Mathematical Circus*, Penguin, UK, 1981

Gardner, Martin, *Mathematical Puzzles and Diversions*, Penguin, UK, 1965

Gibson, Carol (ed.), *Dictionary of Mathematics*, Facts on File, UK, 1988

Goodwin, Matthew, *Numerology: The Complete Guide*, Newcastle Publishing Co., USA, 1981

Gregory, R., *History of Numbers, Time and Money*, Whitcombe & Tombs, New Zealand, 1975

Hogben, Lancelot, *Mathematics for the Million*, Allen & Unwin, UK, 1936

Hood, Peter, *How Time is Measured*, Oxford University Press, UK, 1969

Hunter, John and Cundy, Martyn, *Number*, Blackie, Scotland, UK, 1978

Hurford, James R., *Linguistic Theory of Numerals*, Cambridge University Press, UK, 1975

Irwin, Keith G., *365 Days*, Harrap, UK, 1965

Jordan, Juno and Houston, Helen, *Two Guides to Numerology*, Newcastle Publishing Co., USA, 1982

Kogelman, Stanley, *The Only Maths Book You'll Ever Need*, Facts on File, UK, 1986

Kozminsky, Isidore, *Numbers: Their Meaning and Magic*, Rider, UK, 1985

Le Lionnais, François, *Les Nombres remarquables*, Hermann, France, 1983

*Lines, Malcolm E., *A Number for Your*

Thoughts: Facts and Speculations about Numbers, from Euclid to the Latest Computers, Adam Hilger, UK, 1986

Lunzer, Eric A., *et al.*, *Numbers and the World of Things*, University of Nottingham, School of Education, UK, 1976

Madachy, Joseph S., *Mathematics on Vacation*, Charles Scribner's Sons, USA, 1966

*Menninger, Karl, *Number Words and Number Symbols*, The MIT Press, USA, 1969 (translated from the German)

Miall, Agnes M., *The Book of Fortune Telling*, Treasure Press, UK, 1987

Mordell, Phineas, *The Origin of Letters and Numerals According to the Sefer Yetzirah*, Weiser, USA, 1979

Morris, Desmond, *The Book of Ages*, Jonathan Cape, UK, 1983

Murdin P., and Allen D., *Catalogue of the Universe*, Cambridge University Press, UK, 1979

O'Neil, William M., *Time and the Calendars*, Manchester University Press, UK, 1976

Paling, D. and Fox, J. L., *Numbers and Number Systems*, Oxford University Press, UK, 1968

Parise, Frank (ed.), *The Book of Calendars*, Facts on File, UK, 1982

Parker, Derek and Julia, *How Do You Know Who You Are?*, Thames & Hudson, UK, 1980

Pei, Mario, *The Story of Language*, Allen & Unwin, UK, 1952

Philip, Alexander, *The Calendar: Its History, Structure and Improvement*, Cambridge University Press, UK, 1921

Sampson, Anthony and Sally, *The Oxford Book of Ages*, Oxford University Press, UK, 1985

Sitomer, Mindel and Harry, *How Did Numbers Begin?*, Crowell, USA, 1980

Sondheimer, E., and Rogerson, A., *Numbers and Infinity: A Historical Account of Mathematical Concepts*, Cambridge University Press, UK, 1981

Taylor, Ariel Yvon, *Numerology Made Plain*, Newcastle Publishing Co., USA, 1977

*Urdang, Laurence, *The Facts on File Dictionary of Numerical Allusions*, Facts on File, UK, 1986

*Wells, David, *The Penguin Dictionary of Curious and Interesting Numbers*, Penguin, UK, 1986

Westcott, W. W., *Numbers: Their Occult Power and Mystical Value*, Theosophical Publishing House, UK, 1974

Whitrow, G. J., *What is Time?*, Thames & Hudson, UK, 1972

Withers, Carl, *Counting-Out Rhymes*, Dover, USA, 1970

Young, Michael, *The Metronomic Society: Natural Rhythms and Human Timetables*, Thames & Hudson, UK, 1988

For general literary and traditional references to numbers, the reader is recommended to consult one or more of the following:

Bartlett, John, *Bartlett's Familiar Quotations*, Little, Brown, USA, 1989

Benét, William Rose, *The Reader's Encyclopedia*, A & C Black, UK, 1988

Brewer's Dictionary of Phrase and Fable, revised edition by Ivor H. Evans, Cassell, UK, 1981

The Oxford Dictionary of Quotations, Oxford University Press, UK, 1979

There are also good and detailed entries on such subjects as **numbers** and the **calendar** in the standard encyclopaedias, such as the *Encyclopaedia Britannica*. And most of the major dictionaries give details of the origins of the number-words themselves, and of a number's different meanings and usages.

Finally, mention should specifically be made of the annually published *Whitaker's Almanack*, which has much useful astronomical and chronological information, as well as recent facts, figures and statistical tables, together with tables of a more general nature regarding standard measurements, sizes and the like.

INDEX

(Note: A *Numerical Index* for individual numbers follows this main index on p. 159)

NUMERICAL INDEX